Grollius
Principles of Hydraulics

"If you can`t explain it simply, you don`t understand it well enough."

Albert Einstein

Horst Walter Grollius

Principles of Hydraulics

Second Edition

Univ.-Prof. Dr.-Ing. Horst Walter Grollius
Cologne

Bibliographic information of the German National Library

The German National Library lists this publication in the German National bibliography; detailed bibliographic data are available in the internet: http://dnb.dnb.de

© 2017 Horst Walter Grollius
Production and Publishing:
BoD - Books on Demand, Norderstedt
ISBN: 978-3-7460-3027-2

Preface

To increase the efficiency of production, knowledge and its application in various engineering disciplines is required. This also includes the **fluid technology** which is subdivided in **hydraulics** and **pneumatics**.

With this book the author especially intends to introduce the reader in the principles of hydraulics.

Recourse is made on the book "Grundlagen der Hydraulik" (chapter 2) published by the author in the German language. This book appears in the CARL HANSER-Verlag and is now in the 7th edition.

The book presented here, offers the possibility familiarizing themselves without spending too much time with the **principles of hydraulics**. This particularly applies for students at universities and technical schools. In addition the book will also be of help for those readers which are as technicians in professional practice and want to refresh their basic skills in the field of hydraulics.

In the last chapter the reader will find 10 examples with the detailed presentation of the solution path by the "step by step" method (each step is commented); clarity of the path to find the solution is thus given.

May the study of this book not only make effort, but rather have also motivated the reader to delve with additional literature in this fascinating and economically important field of technology.

Furthermore, many thanks to the company TENADO GmbH (Bochum, Germany); the TENADO CAD software of this company has been used for the creation of all figures shown in the book.

Cologne, November 2017

Horst Walter Grollius

Contents

1 Introduction — 9

2 Physical Principles — 10
    2.1 Pressure Definition, Absolute Pressure, Overpressure. Pressure Units — 10
    2.2 Law of *Pascal* — 12
    2.3 Hydrostatic Pressure — 14
    2.4 Hydraulic Press — 15
    2.5 Pressure Transmission — 17
    2.6 Hydraulic Work, Hydraulik Power, Efficiencies — 18
    2.7 Equation of Continuity — 21
    2.8 *Bernoulli*-Equation — 22
    2.9 Laminar and Turbulent Flows — 24
    2.10 Viscosity — 26
    2.11 Pressure Losses in Pipes, Fittings and Valves — 28
    2.12 Flows through Throttling Devices - Flow Measurement — 36
    2.13 Gap Flows — 39
    2.14 Hydraulic Resistance — 44
    2.15 Compressibility and Compression Module — 47
    2.16 Cavitation — 51

3 Basic Structure of a Hydraulic System — 54

4 Circuit Diagrams — 56

5 Examples — 62
    Example 1: Container with two pistons — 62
    Example 2: Water conducting channel with drain pipe — 63
    Example 3: Pump delivers water from a dam into a container — 67
    Example 4: Oil flows from a tank into a container — 72
    Example 5: Water flows from a reservoir into a channel — 74
    Example 6: Hydraulic press with pressure transmission — 79
    Example 7: Two cylinders which are connected by a pipe — 81
    Example 8: Cylinder to whose piston rod a rope is fastened — 83
    Example 9: An oil filled pipe in different states — 87
    Example 10: Forces acting on piston and piston rod — 91

Sources of Literature — 96

Symbols

Symbols used in the book and not found in the following list will be explained by the book text.

A	Area	m²
B	Width	m
b	Correction factor, gap width	-, m
C	Flow coefficient	-
d	Inner diameter (hydraulic cylinder)	m
dA	Area (infinitesimal small)	m²
dF	Force (infinitesimal small)	N
d_e	Hydaulic diameter	m
d_{PR}	Piston rod diameter	m
E	Modulus of elasticity	N/m²
F	Force	N
F_P	Piston Force	N
G	Weight	N
g	Acceleration of gravity	m/s²
h	Height coordinate, gap height	m, m
I	Electrical current	A
K	True compression module	bar
K_S	Average compression module	bar
k	Absolute wall roughness, correction value	m, -
k/d	Relative pipe roughness	-
l	Pipe length, gap length	m, m
m	Mass	kg
$\dot{m}$	Mass flow	kg/s
P	Hydraulic power	kW
p	Pressure	N/m²
p_{abs}	Absolute pressure	N/m²
p_{amb}	Atmospheric pressure	N/m²
p_e	Overpressure (or gauge pressure)	N/m²
p_I	Inlet pressure (hydraulic pump, hydraulic motor)	N/m²
p_O	Outlet pressure (hydraulic pump, hydraulic motor)	N/m²
Q	Volume flow or flow rate	m³/s

Symbol	Description	Units
R	Spring rate, hydrostatic resistanse	N/m, kg/(m$^4 \cdot$s)
R_{tot}	Total hydrostatic resistance	N/m, kg/(m$^4 \cdot$s)
Re	*Reynolds*-number	-
Re_{crit}	Citical *Reynolds*-number	-
s	Way	m
T	Torque	Nm
t	Time, temperature	s, °C
U	Perimeter, electrical Voltage	m, V
V	Volume	m^3
v	Velocity	m/s
v_m	Average velocity	m/s
v_{max}	Maximum velocity	m/s
v_{Plate}	Plate velocity	m/s
v_{crit}	Critical velocity	m/s
W	Hydraulic work	Nm
β	Ratio of diameters	-
β_P	Isothermal copressibility coefficient	1/bar
Δp	Pressure difference	N/m^2
ς	Flow resistance coefficient	-
η	Dynamic viscosity	N$\cdot$s/m^2

> **NOTE:** For the physical variables used in this book the **International System of Units** (SI) is used. For conversion into units used in Anglo-Saxon countries, conversion tables have to be used, which are available in the web.

1 Introduction

Fluid power is the generic term for the areas of hydraulics and pneumatics. In the area of hydraulics the fluids are liquids; in the area of pneumatics gas is used, namely air. In the beginnings of the hydraulics water was used as the fluid for energy transfer. Since the beginning of the 20th century oils are used. These have lubrication- and corrosion protection in addition. For some years water is also reused as the fluid for energy transfer in individual cases for reasons of environmental protection and costs, also called "water hydraulics". The present book deals mainly with the **physical principals** relevant for oil-operated hydraulic systems (usually mineral oils are used).

The oil-hydraulic is divided into the areas of **hydrodynamic** and **hydrostatic** energy transfer.

The **hydrodynamic energy transfer** uses an impeller in order to transfer mechanical energy to the oil. The flow energy of the oil is used to drive a turbine wheel. These systems are called **hydrodynamic drive** systems (for example Föttinger converters and Fluid couplings).

In the case of the **hydrostatic energy transfer,** a mechanically driven pump (hydraulic pump) produces a mainly pressure-loaded volume flow which is supplied to a hydraulic cylinder or a hydraulic motor. Therein, the pressure energy is reconverted into mechanical energy. These are called **hydrostatic drive** systems.

The kinetic energy is negligible in systems with hydrostatic transfer energy compared to the pressure energy. Conversely, the pressure energy contained in the flow can be neglected in hydrodynamic energy systems. In mechanical engineering, the hydrostatic drive systems have a much greater importance than the hydrodynamic drive systems.

2 Physical Principles

2.1 Pressure Definition, Absolute Pressure, Overpressure, Pressure Units

For the explanation of the **pressure definition** a volume section from a fluid shall be considered as shown in Figure 2.1.

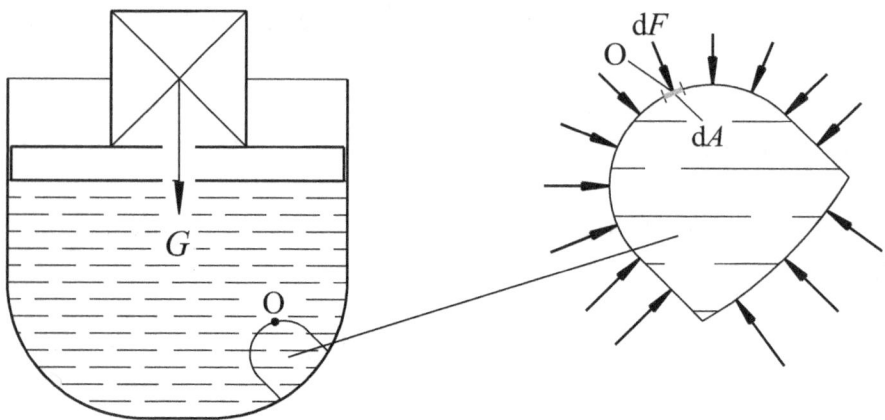

Figure 2.1: For the explanation of the pressure definition

The characteristic fluid point O is equal to a point located on the surface of the part fluid (Figure 2.1, right). At point O the surface element dA is situated, where the force dF is acting vertically. The **pressure** p is the quotient of dF and dA:

$$p = \frac{dF}{dA} \tag{2.1}$$

The pressure value is independent of the cutting sectional plane direction touching point O. That means the pressure is a **scalar** physical quantity; its numerical value depends only on the place in the fluid.

Below, the terms **absolute pressure** and **overpressure** (= pressure measured relative to atmospheric pressure) will be explained based on Figure 2.2.

Physical Principles

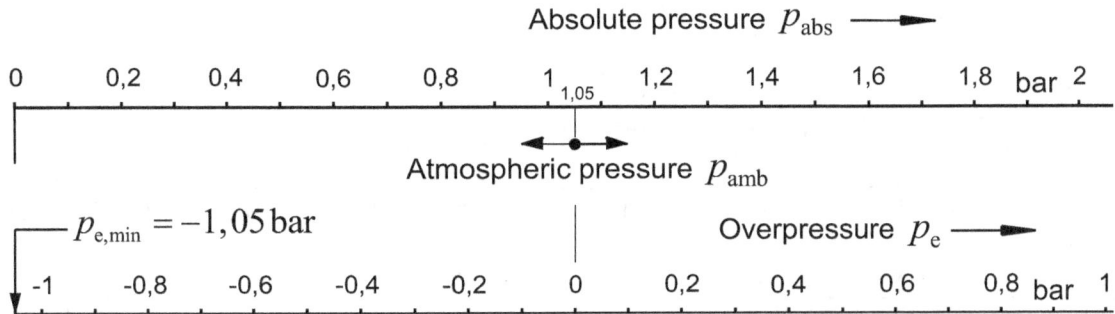

Figure 2.2: Absolute pressure scale and overpressure scale

The **absolute pressure scale** (upper scale in Figure 2.2) starts at $p_{abs}=0$ (pressure at vacuum). The difference between the absolute pressure p_{abs} and the local (absolute) atmospheric pressure p_{amb} is the atmospheric pressure difference:

$$p_e = p_{abs} - p_{amb} \tag{2.2}$$

This pressure difference is called **overpressure** (or **gauge pressure**).

If the absolute pressure p_{abs} is higher than the local (absolute) atmospheric pressure p_{amb} the overpressure became positive value

$$p_e = p_{abs} - p_{amb} > 0 \tag{2.3}$$

If the absolut pressure p_{abs} is lower than the actual (absolute) atmospheric pressure p_{amb} the overpressure became negative value

$$p_e = p_{abs} - p_{amb} < 0 \tag{2.4}$$

The minimal (theoretical) overpressure value $p_{e,min}$ is determined by the actual (absolute) atmospheric pressure p_{amb}. For example, if there is a pressure with $p_{amb}=1,05\,bar$ as shown in Figure 2.2 the minimal overpressure value is

$$p_{e,min} = p_{abs,min} - p_{amb} = 0\,\text{bar} - 1,05\,\text{bar} = -1,05\,\text{bar} \tag{2.5}$$

The example shows: The numerical value of the minimal overpressure value is depending on the actual (absolute) atmospheric pressure value p_{amb}.

> NOTE: Often the indices "abs" and "e" are omitted for clear identification of absolute pressure and overpressure. From the context it is to find out whether absolute pressure or overpressure is of importance.

A commonly used **unit of pressure** based on the **International SI-System** is **Pascal** (unit symbol: Pa)

$$1\,\text{Pa} = 1\,\frac{\text{N}}{\text{m}^2} = 1\,\frac{\frac{\text{kg}}{\text{m}\cdot\text{s}^2}}{\text{m}^2} \tag{2.6}$$

(Pa = Pascal, N = Newton, kg = kilogram, m = meter, s = second)

An also often used unit is **Bar** (unit symbol: bar):

$$1\,\text{bar} = 10^5\,\text{Pa} = 10^5\,\frac{\text{N}}{\text{m}^2} = 10^5\,\frac{\frac{\text{kg}}{\text{m}\cdot\text{s}^2}}{\text{m}^2} \tag{2.7}$$

Small pressure values are given in **millibar** (unit symbol: mbar) or **hectopascal** (unit symbol: hPa)

$$1\,\text{mbar} = 0,001\,\text{bar} = 1\,\text{hPa} \tag{2.8}$$

The unit used in Anglo-Saxon countries is **Psi** (unit symbol: psi):

$$1\,\text{bar} = 14,50377\,\text{psi} \approx 14,5\,\text{psi} \tag{2.9}$$

2.2 Law of *Pascal*

The law of *Pascal* is the **fundamental law of hydrostatics**. It is valid for incompressible liquids. The effect of gravity is ignored. It states the following

Physical Principles

If a liquid in a container is influenced by a pressure at any place (for example by a force loaded piston) thus the pressure on the inner wall of the container and inside the liquid has the same numeric value.

For a better understanding of the law of *Pascal* is to look at Figure 2.3.

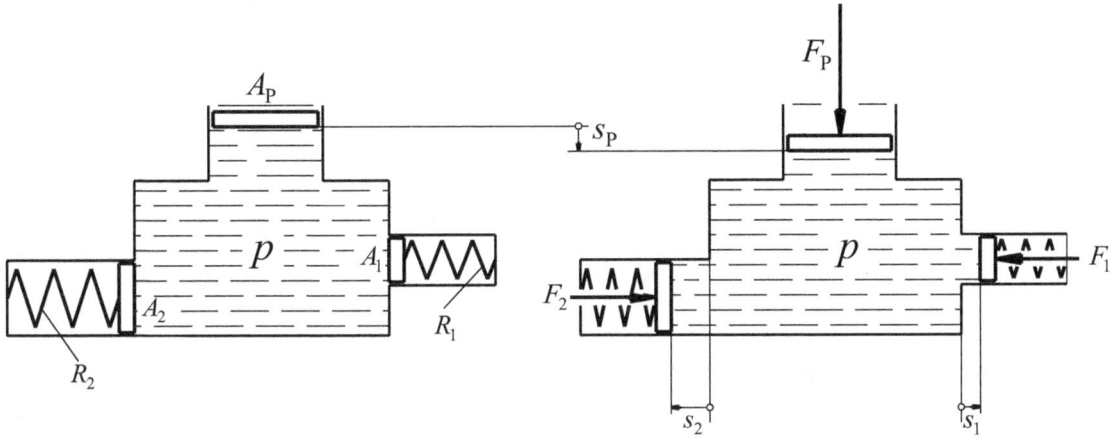

Figure 2.3: For the explanation of the law of *Pascal*

With the movement of the piston around the way s_P down the fluid volume $V_P = A_P \cdot s_P$ is displaced. This volume finds place in both lateral chambers, which are sealed by freely of friction controlled pistons without leakage. It is

$$V_P = V_1 + V_2 = s_1 \cdot A_1 + s_2 \cdot A_2 \qquad (2.10)$$

Due to the movement of the pistons in the chambers, at the side arranged compression springs are pressed. This entails that the spring forces F_1 and F_2 are acting about their respective piston surface on the liquid. From the right piston acts on the liquid the pressure

$$p_1 = \frac{F_1}{A_1} = \frac{R_1 \cdot s_1}{A_1} \qquad (2.11)$$

From the left piston acts on the liquid the pressure

$$p_2 = \frac{F_2}{A_2} = \frac{R_2 \cdot s_2}{A_2} \tag{2.12}$$

Assuming the spring ways s_1, s_2 and the spring rates R_1, R_2 and the piston areas A_1, A_2 are known, the calculation of the pressures gives

$$p_1 = p_2 \tag{2.13}$$

The **law of *Pascal*** is therefore confirmed.

The **pressure** in the container is generally named p. That gives

$$p = p_1 = p_2 = p_P \tag{2.14}$$

The force acting on the upper piston in its final position is

$$F_P = p \cdot A_P \tag{2.15}$$

2.3 Hydrostatic Pressure

The law of *Pascal* applies under the assumption that the effect of gravity is ignored. There is no influence of gravity on the fluid in the container: the fluid will be considered weightless. Nevertheless, in reality the fluid is under the influence of gravity and beside the pressure generated by external forces the pressure caused by the gravity, the so-called **hydrostatic pressure**, still exists. Figure 2.4 shows a fluid-filled container which is open at the top. On the liquid level at $h = 0$ the atmospheric pressure p_{amb} has an effect. The graph beside the container illustrates the pressure curve in the fluid in response to the height coordinate h.

Physical Principles

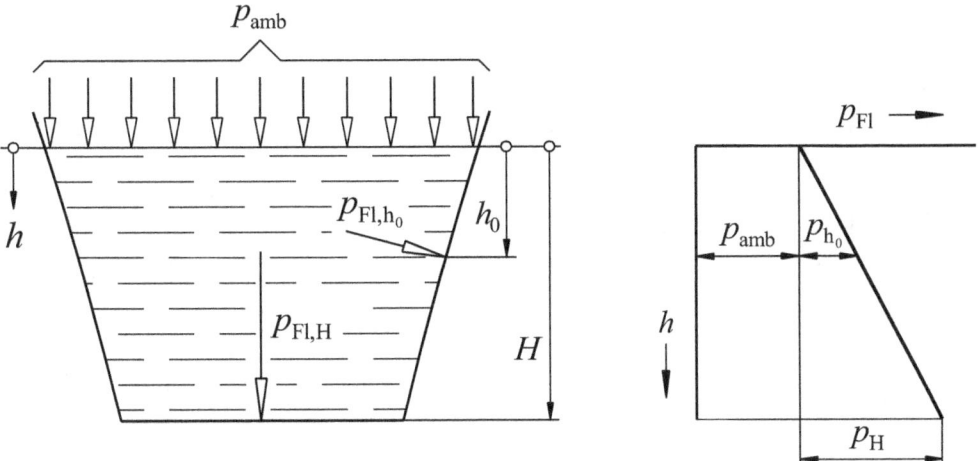

Figure 2.4: For explaining the hydrostatic pressure in a fluid

Only from the gravity generated pressure in the fluid is given by

$$\boxed{p_h = \rho \cdot g \cdot h} \tag{2.16}$$

For the pressure in the fluid in taking account the atmospheric pressure in the depth h_0 we obtain

$$p_{Fl,h_0} = p_{amb} + p_{h_0} = p_{amb} + \rho \cdot g \cdot h_0 \tag{2.17}$$

On the container base H the pressure acts

$$p_{Fl,H} = p_{amb} + p_H = p_{amb} + \rho \cdot g \cdot H \tag{2.18}$$

> NOTE: During the design and calculation of hydraulic systems is to be checked whether the hydrostatic pressure accepts a notable size compared with pressures appearing in the system (**system pressures**). Mostly the hydrostatic pressure finds no consideration, because this pressure often is negligible low compared with the system pressures.

2.4 Hydraulic Press

The fundamental functionality of the hydraulic press should be explained on the basis of Figure 2.5.

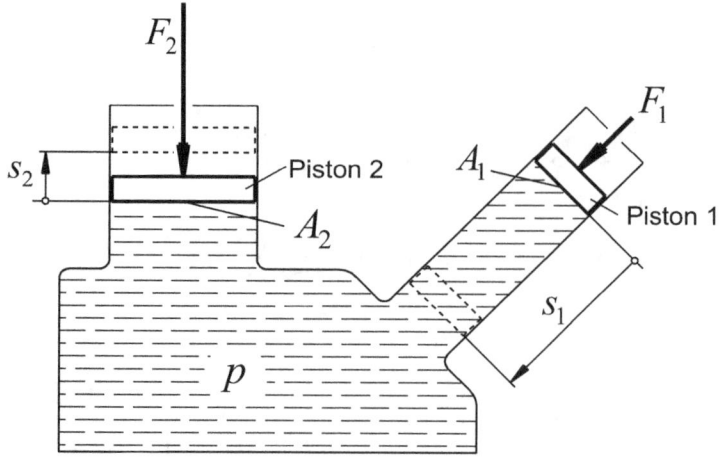

Figure 2.5: For explaining the hydraulic press

The influence of the hydrostatic pressure remains disregarded. The pistons of the hydraulic press are sealed by freely of friction controlled pistons without leakage.

$$p = \frac{F_1}{A_1} \qquad (2.19)$$

The pressure p acts according to the law of *Pascal* on all places of the fluid. Therefore the pressure p acts also on the area A_2 of the piston 2. With

$$p = \frac{F_1}{A_1} = \frac{F_2}{A_2} \qquad (2.20)$$

we obtain

$$F_2 = F_1 \frac{A_2}{A_1} \qquad (2.21)$$

With equation (2.21) the principle of **force transmission** can be made clear. For example: $A_2 = 10 \cdot A_1 \quad \Rightarrow \quad F_2 = 10 \cdot F_1$

Physical Principles 17

With the movement of the piston 1 around the way s_1 the volume $V_1 = A_1 \cdot s_1$ is displaced. The piston 2 is thereby moved around the way s_2 upwards. It is

$$V_1 = A_1 \cdot s_1 = A_2 \cdot s_2 \tag{2.22}$$

We obtain

$$\boxed{s_2 = s_1 \frac{A_1}{A_2}} \tag{2.23}$$

With equation (2.23) the principle of **way transmission** can be made clear. For example: $A_2 = 10 \cdot A_1 \quad \Rightarrow$

$$s_2 = \frac{A_1}{A_2} s_1 = \frac{A_1}{10 \cdot A_1} s_1 = \frac{1}{10} s_1 \tag{2.24}$$

2.5   Pressure Transmission

The principle of pressure transmission should be explained on the basis of Figure 2.6.

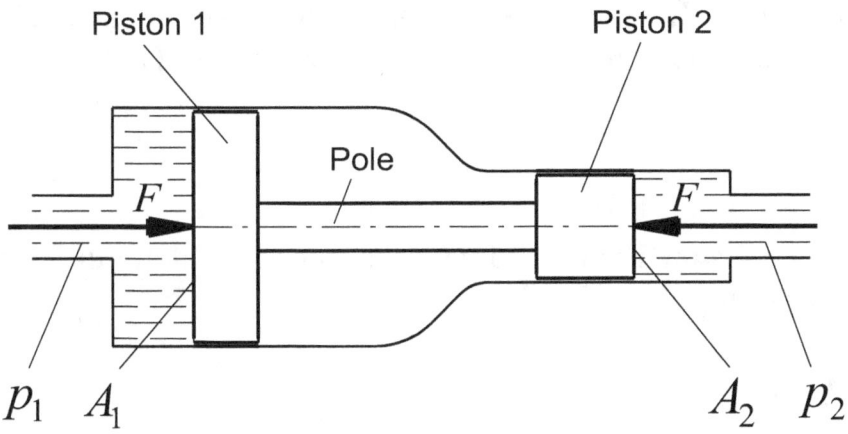

Figure 2.6: For the explanation of the pressure transmission

Both freely of friction controlled and without leakage sealed pistons (areas A_1 and A_2) are connected by a pole firmly with each other.

The pressure p_1 acts at the surface A_1. The piston 1 attacking force is therefore $F = p_1 \cdot A_1$. For reasons of the static balance the force F attacks the piston 2 too. The pressure at the piston surface A_2 is therefore $p_2 = F/A_2$. It is

$$F = p_1 \cdot A_1 = p_2 \cdot A_2 \tag{2.25}$$

So we get

$$\boxed{p_2 = p_1 \frac{A_1}{A_2}}$$
(2.26)

With equation (2.26) the principle of **pressure transmission** can be made clear. For example: $A_1 = 2 \cdot A_2 \quad \Rightarrow \quad p_2 = 2 \cdot p_1$

2.6 Hydraulic Work, Hydraulic Power, Efficiencies

For explaining the term **hydraulic work** Figure 2.5 has to be looked. The piston 1 is moved with the force F_1 along the way s_1. It is done the hydraulic work

$$W_1 = F_1 \cdot s_1 = p_1 \cdot A_1 \cdot s_1 \tag{2.27}$$

During this process the piston 2 is moved with the force F_2 along the way s_2. It is done the hydraulic work

$$W_2 = F_2 \cdot s_2 = p_2 \cdot A_2 \cdot s_2 \tag{2.28}$$

Under use of $V_1 = A_1 \cdot s_1$ and $V_2 = A_2 \cdot s_2$ we get

Physical Principles

$$\boxed{W_1 = p_1 \cdot V_1} \quad \text{and} \quad \boxed{W_2 = p_2 \cdot V_2} \qquad (2.29), (2.30)$$

The **hydraulic power** P_1 is the quotient from the hydraulic work W_1 and the time t_1, which is required to move the piston 1 around the way s_1.

$$P_1 = \frac{W_1}{t_1} = \frac{p_1 \cdot V_1}{t_1} \qquad (2.31)$$

With the flow rates $Q_1 = V_1 / t_1$ we get

$$\boxed{P_1 = p_1 \cdot Q_1}$$
(2.32)

In analogous way we receive the **hydraulic power** P_2 for piston 2

$$\boxed{P_2 = p_2 \cdot Q_2} \qquad (2.33)$$

The **total efficiency** of a hydraulic pump and a hydraulic motor is given by the equation

$$\eta_t = \eta_v \cdot \eta_{hm} \qquad (2.34)$$

In equation (2.34) η_v means **volumetric efficiency**. It takes into account the so called **volumetric losses** caused by leakages. The **hydraulic-mechanical efficiency** η_{hm} is a measure for the losses that caused by **flow losses** and **friction losses**. Friction losses are the losses caused by each other gliding machine parts. Figure 2.7 is intended to illustrate the term of **total efficiency**.

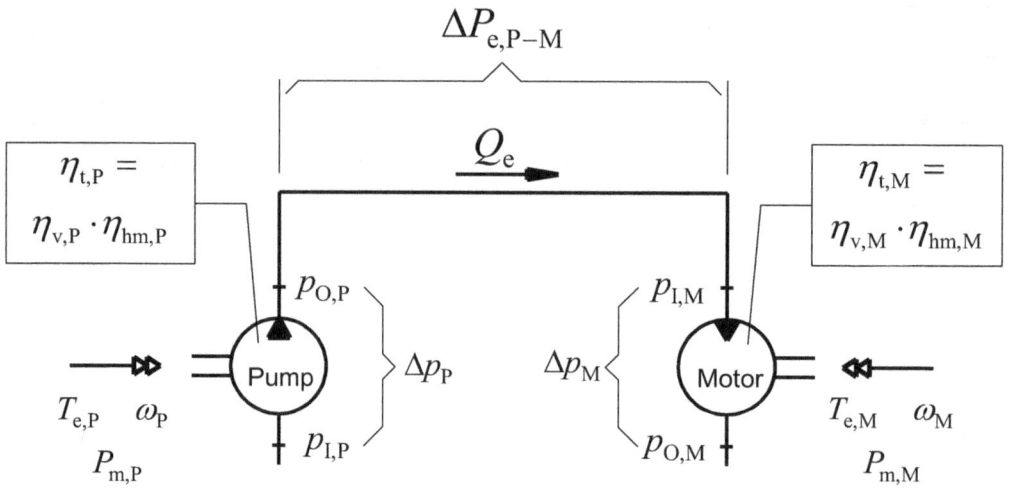

Figure 2.7: For illustrating the term of total efficiency

The shaft power of the hydraulic pump (mechanical input power) is calculated by using the equation $P_{m,P} = T_{e,P} \cdot \omega_P$. This power is converted to a large extent in the hydraulic power $P_{e,P} = \Delta p_P \cdot Q_e$. A small portion of the shaft power is needed to cover the volumetric losses caused by leakages and to cover the flow and the friction losses, so that results $P_{e,P} < P_{m,P}$.

The **total efficiency** of the hydraulic pump is

$$\eta_{t,P} = \frac{P_{e,P}}{P_{m,P}} = \frac{\Delta p_P \cdot Q_e}{T_{e,P} \cdot \omega_P} = \frac{(p_{O,P} - p_{I,P}) Q_e}{T_{e,P} \cdot \omega_P} \qquad (2.35)$$

The power available to the hydraulic motor $P_{e,M}$ is due to the occurring power losses $\Delta P_{e,P-M}$ between the outlet connection of the hydraulic pump and the inlet connection of the hydraulic motor smaller than the power present at the outlet of the pump. This is the reason that results

$$P_{e,M} = P_{e,P} - \Delta P_{e,P-M} \qquad (2.36)$$

The hydraulic power $P_{e,M}$ is for the most part available on the shaft of the motor in the form of mechanical power $P_{m,M} = T_{e,M} \cdot \omega_M$. Also in the hydraulic motor occur volumetric losses, flow losses and friction losses. These are covered partly by the hydraulic power, so that results $P_{m,M} < P_{e,M}$.

The **total efficiency** of the hydraulic motor is

$$\eta_{t,M} = \frac{P_{m,M}}{P_{e,M}} = \frac{T_{e,M} \cdot \omega_M}{\Delta p_M \cdot Q_e} = \frac{T_{e,M} \cdot \omega_M}{(p_{I,M} - p_{O,M})Q_e} \qquad (2.37)$$

2.7 Equation of Continuity

According to Figure 2.8 a fluid flows through a pipe with varying cross-sectional areas.

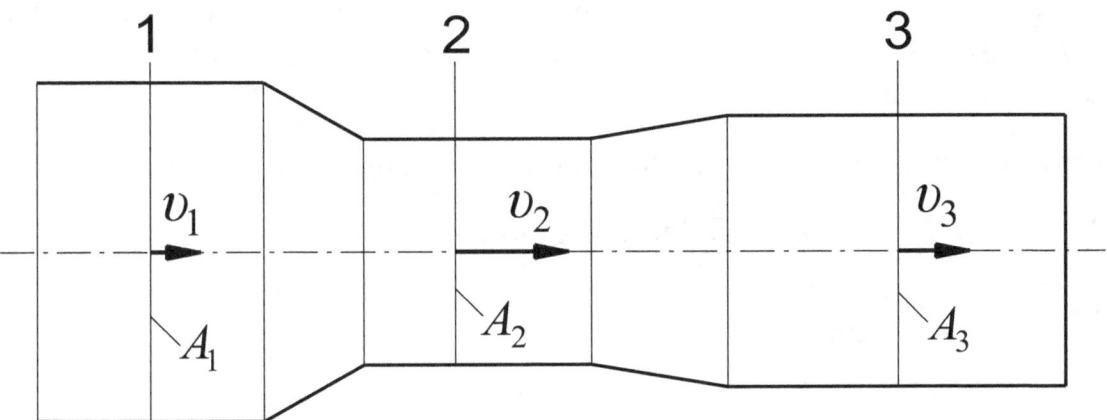

Figure 2.8: Constancy of the volume flow - incompressible fluid

Between the 1, 2 and 3 marked cross-sectional areas occurs no loss of fluid. Therefore results to the mass flows flowing through these areas

$$m_1 = m_2 = m_3 \qquad (2.38)$$

$m_1 = Q_1 \cdot \rho_1 = A_1 \cdot v_1 \cdot \rho_1 \qquad m_2 = Q_2 \cdot \rho_2 = A_2 \cdot v_2 \cdot \rho_2 \qquad m_3 = Q_3 \cdot \rho_3 = A_3 \cdot v_3 \cdot \rho_3$

$$A_1 \cdot v_1 \cdot \rho_1 = A_2 \cdot v_2 \cdot \rho_2 = A_3 \cdot v_3 \cdot \rho_3 \tag{2.39}$$

Liquids - included oils used in hydraulic - can be compressed only slightly. That is why results

$$\rho_1 \approx \rho_2 \approx \rho_3 \tag{2.40}$$

That gives with this assumption the **equation of continuity**

$$\boxed{Q_1 = A_1 \cdot v_1 = Q_2 = A_2 \cdot v_2 = Q_3 = A_3 \cdot v_3} \tag{2.41}$$

2.8 *Bernoulli*-Equation

The *Bernoulli*-equation represents a special case of the *Navier-Stokes* equations known from fluid mechanics. These equations are valid for three-dimensional viscosity-afflicted flows. If we assume the flow is steady, frictionless (without losses), incompressible and one-dimensional the *Navier-Stokes* equations simplify themselves to the equation named *Bernoulli*-equation.

Flows are **stationary** if their variables do not change with the time.

The *Bernoulli*-equation is in the **energy-form**

$$\boxed{\frac{v^2}{2} + g \cdot z + \frac{p}{\rho} = \text{const.}} \tag{2.42}$$

This equation shows: The total energy as the sum of kinetic energy $v^2/2$, potential energy $g \cdot z$ and pressure energy p/ρ is constant.

The *Bernoulli*-equation is in the **height-form**

$$\boxed{\frac{v^2}{2 \cdot g} + z + \frac{p}{\rho \cdot g} = \text{const.}} \tag{2.43}$$

Physical Principles

The *Bernoulli*-equation is in the **pressure-form**

$$\boxed{\rho \frac{v^2}{2} + \rho \cdot g \cdot z + p = \text{const.}} \tag{2.44}$$

Figure 2.9 illustrates the application of equation (2.43).

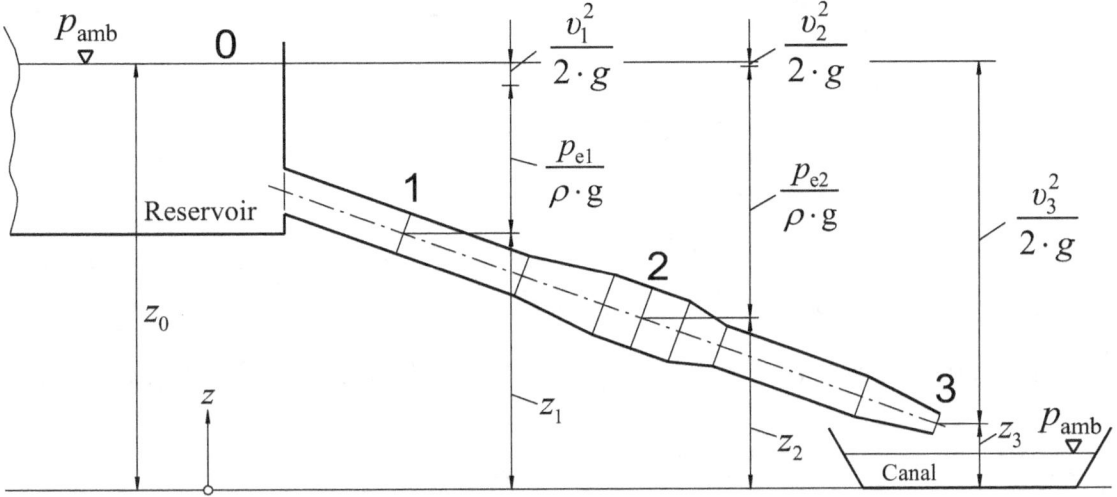

Figure 2.9: For illustrating equation (2.43)

For the upper water level **0** and the pipe cross sections **1, 2** and **3** can be formulated ($\rho_0 = \rho_1 = \rho_2 = \rho_3 = \rho$)

$$z_0 + \frac{p_0}{\rho \cdot g} + \frac{v_0^2}{2 \cdot g} = z_1 + \frac{p_1}{\rho \cdot g} + \frac{v_1^2}{2 \cdot g} =$$

$$z_2 + \frac{p_2}{\rho \cdot g} + \frac{v_2^2}{2 \cdot g} = z_3 + \frac{p_3}{\rho \cdot g} + \frac{v_3^2}{2 \cdot g} \tag{2.45}$$

The absolute pressures p_0, p_1, p_2 and p_3 are now replaced by the sum of the atmospheric pressure and the respective overpressure

$$z_0 + \frac{p_{amb} + p_{e0}}{\rho \cdot g} + \frac{v_0^2}{2 \cdot g} = z_1 + \frac{p_{amb} + p_{e1}}{\rho \cdot g} + \frac{v_1^2}{2 \cdot g} =$$

$$z_2 + \frac{P_{amb} + P_{e2}}{\rho \cdot g} + \frac{v_2^2}{2 \cdot g} = z_3 + \frac{P_{amb} + P_{e3}}{\rho \cdot g} + \frac{v_3^2}{2 \cdot g} \qquad (2.46)$$

$$z_0 + \frac{P_{amb}}{\rho \cdot g} + \frac{P_{e0}}{\rho \cdot g} + \frac{v_0^2}{2 \cdot g} = z_1 + \frac{P_{amb}}{\rho \cdot g} + \frac{P_{e1}}{\rho \cdot g} + \frac{v_1^2}{2 \cdot g} =$$

$$z_2 + \frac{P_{amb}}{\rho \cdot g} + \frac{P_{e2}}{\rho \cdot g} + \frac{v_2^2}{2 \cdot g} = z_3 + \frac{P_{amb}}{\rho \cdot g} + \frac{P_{e3}}{\rho \cdot g} + \frac{v_3^2}{2 \cdot g} \qquad (2.47)$$

$$z_0 + \frac{P_{e0}}{\rho \cdot g} + \frac{v_0^2}{2 \cdot g} = z_1 + \frac{P_{e1}}{\rho \cdot g} + \frac{v_1^2}{2 \cdot g} =$$

$$z_2 + \frac{P_{e2}}{\rho \cdot g} + \frac{v_2^2}{2 \cdot g} = z_3 + \frac{P_{e3}}{\rho \cdot g} + \frac{v_3^2}{2 \cdot g} \qquad (2.48)$$

It is still considered that is valid $p_{e0} = 0$, $p_{e3} = 0$ and $v_0 = 0$, then the result is

$$\boxed{z_0 = z_1 + \frac{P_{e1}}{\rho \cdot g} + \frac{v_1^2}{2 \cdot g} = z_2 + \frac{P_{e2}}{\rho \cdot g} + \frac{v_2^2}{2 \cdot g} = z_3 + \frac{v_3^2}{2 \cdot g}} \qquad (2.49)$$

NOTE: The system shown in Figure 2.9 is also used in the example 5 that illustrates the equation (2.49) using numerical values.

2.9 Laminar and Turbulent Flows

In the hydraulic piping systems occur laminar or turbulent flows.

In the **laminar flow** (Figure 2.10, left), the fluid particles move in orderly separate layers. The flow lines are parallel to the pipe axis. This character of a laminar flow can be detected for example by introducing a thin coloured liquid jet in a through-flow of water pipe: the liquid jet is preserved in form and colour without interfering with the surrounding water. The maximum value of the flow velocity is in the middle of the pipe; about the pipe cross section a parabolic velocity profile arises. In the **turbulent flow** (Figure 2.10, right) the fluid particles do not move in orderly layers as in the laminar flow. The axially - in the direction of the tube axis - running main flow now overlap at all places randomly occurring longitudinal and transverse movements. The flow is therefore more or less mixed. About the pipe cross section a nearly constant velocity profile arises, which drops sharply toward the pipe wall. Near the pipe wall there is a thin layer with the thickness δ in which the flow is laminar (laminar layer).

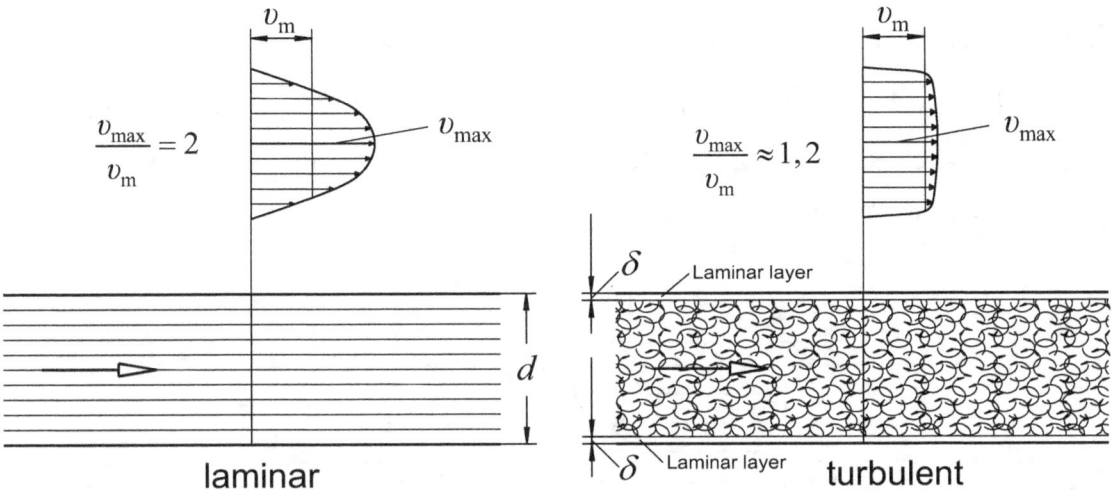

Figure 2.10: Main differences between laminar and turbulent flows

By means of a number named to *Reynolds* can be checked which flow form - laminar or turbulent - is given in a straight pipe with circular cross section. This is the **Reynolds-number** Re

$$Re = \frac{v \cdot d}{\nu} \qquad (2.50)$$

In equation (2.50) means v the flow velocity (average value), d means the inside diameter of the pipe and ν means the **kinematic viscosity** of the fluid.

Note: In section 2.10 there is presented the definition of kinematic viscosity.

The change from laminar to turbulent flow occurs in straight pipes with circular cross section at the **critical Reynolds-number** $Re_{krit} = 2320$.

From equation (2.50) can be calculated under use from Re_{krit} the **critical velocity** v_{krit} at which the change from laminar to turbulent flow occurs

$$v_{krit} = \frac{Re_{krit} \cdot v}{d} = \frac{2320 \cdot v}{d} \tag{2.51}$$

2.10 Viscosity

According to Figure 2.11 a plate having the area A lies on a liquid layer with the thickness h. The plate is moved at constant velocity v_{Plate} in parallel with a standing still wall ($v = 0$).

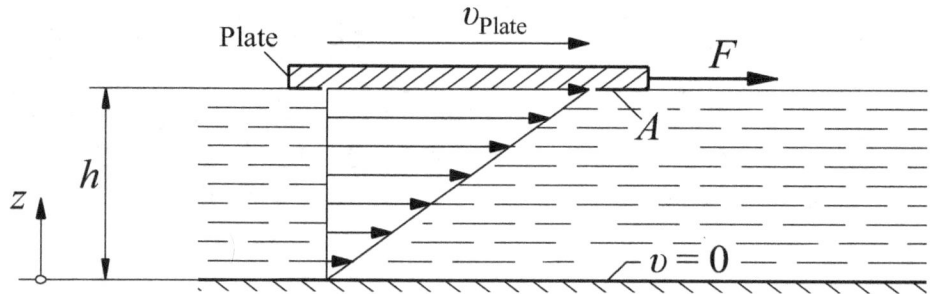

Figure 2.11: To the Newtonian law of friction

In order to maintain the movement the force F is required. Between the plate and the standing still wall a linear velocity gradient is formed. However, this is valid only if the thickness h of the liquid layer is not too big. The following law discovered by *Newton* is known as the **Newtonian law of friction**.

$$\frac{F}{A} = \tau = \eta \frac{dv}{dz} \tag{2.52}$$

It mean: τ the **friction shear stress** (unit: N/m^2) and η the **dynamic viscosity** of the liquid (unit: $N \cdot s/m^2$). The dynamic viscosity η can be regarded as a measure for the friction-work which is caused by the liquid-

Physical Principles

particles when slide against each other; the work expended is converted into heat. The definition of the **kinematic viscosity** also used in hydraulics is

$$\nu = \frac{\eta}{\rho}$$ (2.53)

The unit of ν is m^2/s. To calculate the **Reynolds-number** Re, ν is required, see equation (2.50).

Figure 2.12 shows exemplarily the typical viscosity-temperature-pressure behaviour of hydraulic oils.

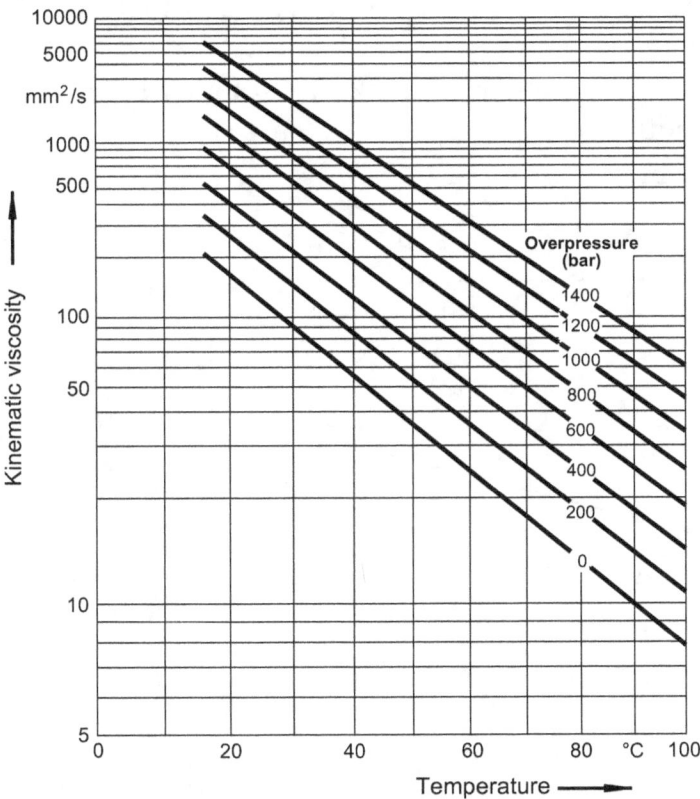

Figure 2.12: Typical viscosity-temperature-pressure behaviour (*Shell*)

2.11 Pressure Losses in Pipes, Fittings and Valves

In the frictionless flow of a liquid, the total energy (flow energy) as the sum of pressure energy, kinetic energy and potential energy is constant, expressed by the *Bernoulli*-equation (2.42).

In the flow of a liquid with friction (real flow) a part of the flow energy is converted because of the influence of the viscosity into heat. This thermal energy cannot be used technically and therefore is designated as loss energy or **flow loss**. The potential energy and the kinetic energy cannot be affected by loss of energy, because the local heights are not altered by the friction and the flow velocities are predetermined by the equation of continuity. Of losses due to friction influences only the pressure energy can therefore be affected.

For a pipe with a constant cross section (Figure 2.13) through which a real liquid flows, the following applies.

$$\rho \frac{v_1^2}{2} + \rho \cdot g \cdot z_1 + p_1 = \rho \frac{v_2^2}{2} + \rho \cdot g \cdot z_2 + p_2 + \Delta p_R \qquad (2.54)$$

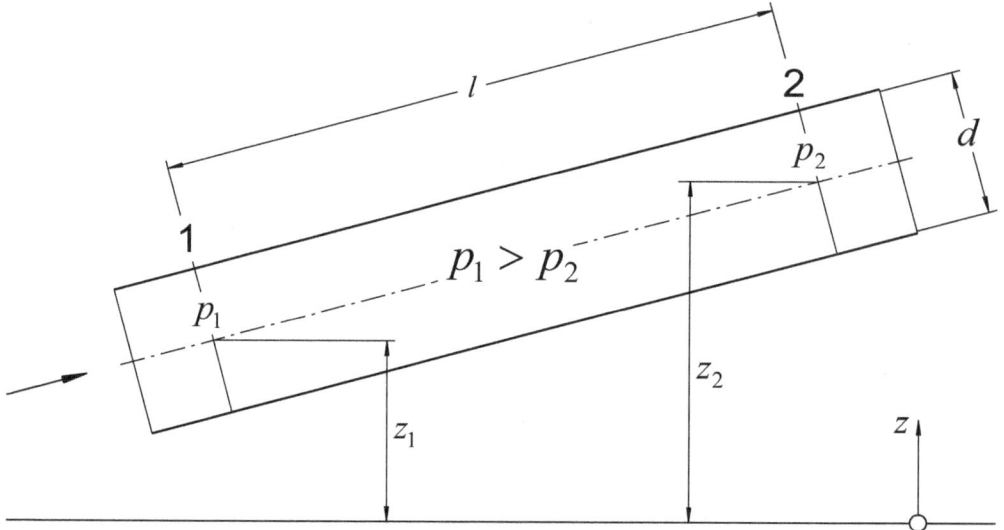

Figure 2.13: To the pressure loss in a pipe

In the equation (2.54) Δp_R is the pressure loss caused by friction influence.

Prandtl has established for **incompressible, stationary** and **isothermal** flows the equation

$$\Delta p_R = \lambda_R \frac{l}{d} \frac{\rho \cdot v^2}{2} \qquad (2.55)$$

In it is λ_R the **pipe friction coefficient**. Rearranging the equation (2.54) gives

$$p_1 - p_2 = \frac{\rho}{2}\left(v_2^2 - v_1^2\right) + \rho \cdot g\left(z_2 - z_1\right) + \Delta p_R \qquad (2.56)$$

For the pipe with a constant cross section over the length l shown in Figure 2.13 results according to the continuity equation

$$v_1 = v_2 \qquad (2.57)$$

This is according to equation (2.56)

$$p_1 - p_2 = \rho \cdot g\left(z_2 - z_1\right) + \Delta p_R \qquad (2.58)$$

Equation (2.58) shows that the pressure difference $p_1 - p_2$ between two flow cross sections in a pipe with constant cross section over the length l is determined by the local heights z_1 and z_2, the density of the liquid ρ, the acceleration due to gravity g and the pressure loss Δp_R caused by the influence of friction.

In a horizontally installed pipe ($z_1 = z_2$) with constant cross section over the length l is in accordance with equation (2.58) the pressure difference $p_1 - p_2$ only caused by friction influences. It is

$$p_1 - p_2 = \Delta p_R = \lambda_R \frac{l}{d} \frac{\rho \cdot v^2}{2} \qquad (2.59)$$

For normally in the hydraulic occurring *Reynolds*-numbers which are in the range $600 < Re < 60000$ one uses to determine the pipe friction coefficient the chart shown in Figure 2.14 and the equations that can be found in there.

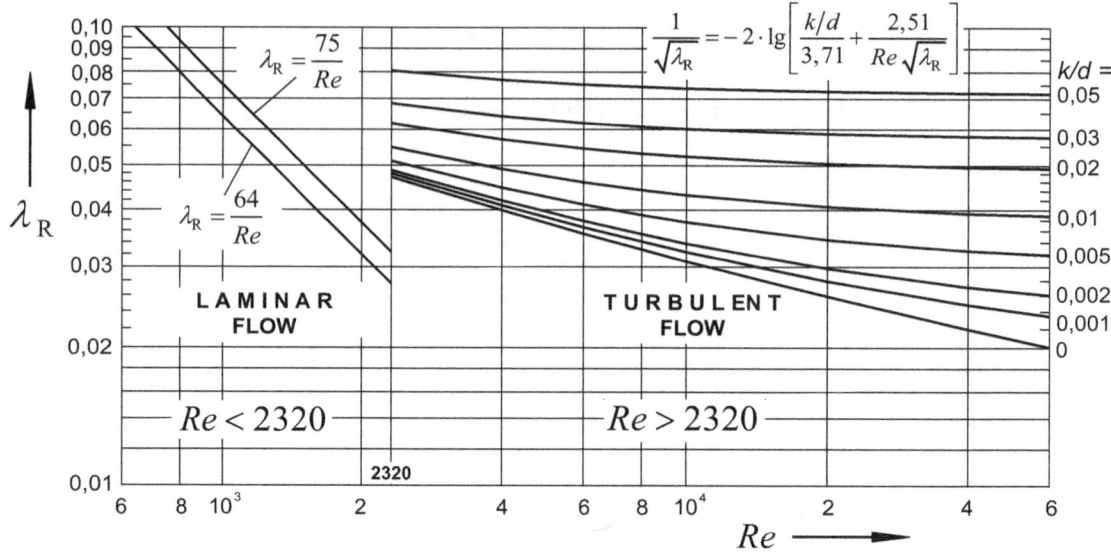

Figure 2.14: Chart to determine the pipe friction coefficient

Mostly in **laminar flow** the following equation is used to calculate the pipe friction coefficient

$$\lambda_R = \frac{64}{Re} \tag{2.60}$$

Equation (2.60) is valid under the condition that the flow **isothermal** runs. This means that the flow temperature neither increases nor decreases ($T =$ const.).

Fluids flow through pipes of hydraulic systems, as well as zones of heating and cooling may be present ($T \neq$ const.). In such cases, we get more realistic results with equation

$$\lambda_R = \frac{75}{Re} \tag{2.61}$$

Physical Principles

This equation comes from *Panzer* and *Beitler*.

If the **flow is turbulent**, the following equation offers the possibility of calculating the pipe friction coefficient λ_R to a **good approximation**

$$\frac{1}{\sqrt{\lambda_R}} = -2 \lg\left[\frac{k/d}{3,71} + \frac{2,51}{Re\sqrt{\lambda_R}}\right] \qquad (2.62)$$

Equation (2.62) comes from *Colebrook* and is based on formulas that *Prandtl*, *von Kármán* and *Nikuradse* have developed on the basis of extensive studies. As we can see, λ_R is dependent on the **Reynolds-number** Re and the **relative pipe-roughness** k/d; k is known as the **absolute wall roughness**. It is a measure of the roughness of the pipe wall surfaces.

For practical use equation (2.62) has the disadvantage that the desired physical variable λ_R is implicit in the equation. It is recommended to design a mathematical algorithm that can be solved programmed using a PC. Values for the wall roughness are found in *Wärmetechnische Arbeitsmappe*.

> NOTE: If the *Reynolds*-number lies close to Re_{krit}, the here presented equations may not provide realistic values. It is therefore recommended, in particular by the choice of the inner pipe diameter to influence the *Reynolds*-number in order to get a suitable distance to Re_{krit}.

In pipes and hoses of hydraulic systems, the (average) flow velocity should not exceed the value of 10 m/s if possible.

The calculation of the pressure loss with equation (2.55) requires a numerical value for the size of d. For pipes with circular cross section, this is set equal to the inner diameter.

For pipes with non-circular cross-section (for example tubes with square or rectangular cross section) by means of the following equation calculated diameter is used

$$d_e = 4\left(\frac{A}{U}\right)_{nc} \tag{2.63}$$

This diameter is called **replacement diameter** or **hydraulic diameter**. In equation (2.63) A is the cross-sectional area and U is the wetted perimeter of the flow-through of non-circular cross-sectional pipe.

For tubes with constant cross-section over the length (however, other than circular) in such cases for the calculation of the pressure loss, we use the equation

$$\Delta p_R = \lambda_R \frac{l}{d_e} \frac{\rho \cdot v^2}{2} = \lambda_R \frac{l}{4\left(\frac{A}{U}\right)_{nc}} \frac{\rho \cdot v^2}{2} \tag{2.64}$$

For the determination of λ_R, the **Reynolds-number** Re is required. This is calculated in pipes with non-circular cross sections by using the following equation

$$Re = \frac{v \cdot d_e}{v} = 4 \frac{v \left(\frac{A}{U}\right)_{nc}}{v} \tag{2.65}$$

Influences of friction can cause large pressure losses in **pipe fittings** (e.g. pipe bends, pipe branches, extensions, restrictions). Their calculation is done using a size determined by experiments, the **flow resistance coefficient** ς

$$\Delta p_F = \varsigma \frac{\rho \cdot v^2}{2} \tag{2.66}$$

Figure 2.15 shows flow resistance coefficients ς for pipe branches according to *Chaimowitsch*. These are valid for **turbulent flow**.

Physical Principles

The correction factor b takes into account the increase of the pressure loss with decreasing of the *Reynolds*-number.

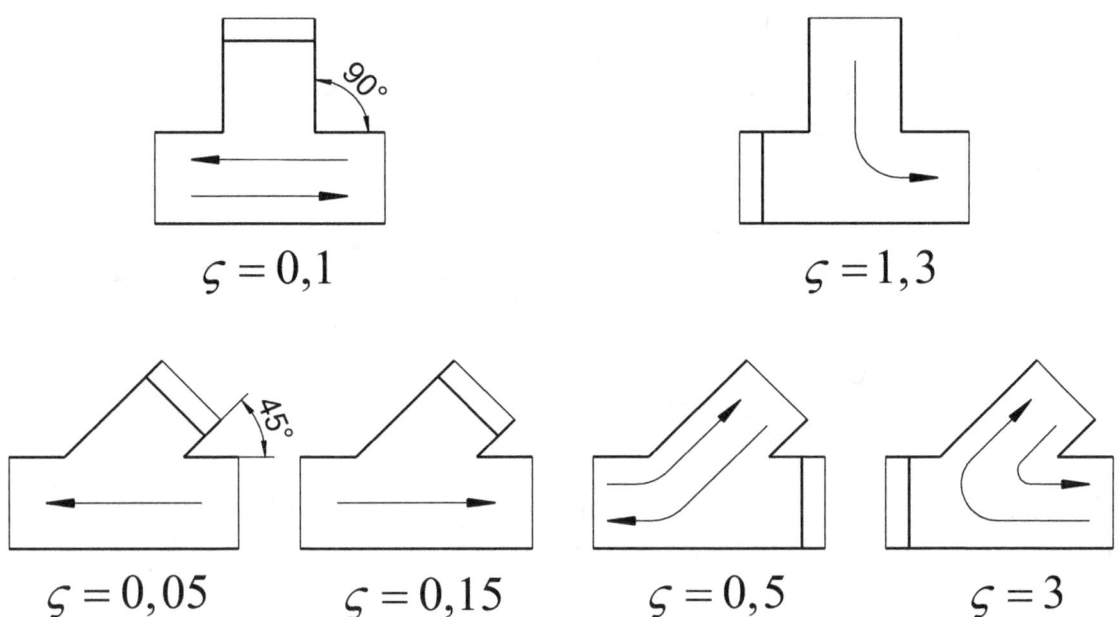

Figure 2.15: Selection of flow resistance coefficients for 90°- and 45°-pipe elbows (*Chaimowitsch*)

At low *Reynolds*-number, the pressure loss Δp_F is calculated based on an equation taking into account a correction factor b

$$\Delta p_F = \varsigma \frac{\rho \cdot v^2}{2} b \qquad (2.67)$$

Figure 2.16 shows the dependency of the correction factor b of the *Reynolds*-number Re according to *Chaimowitsch*.

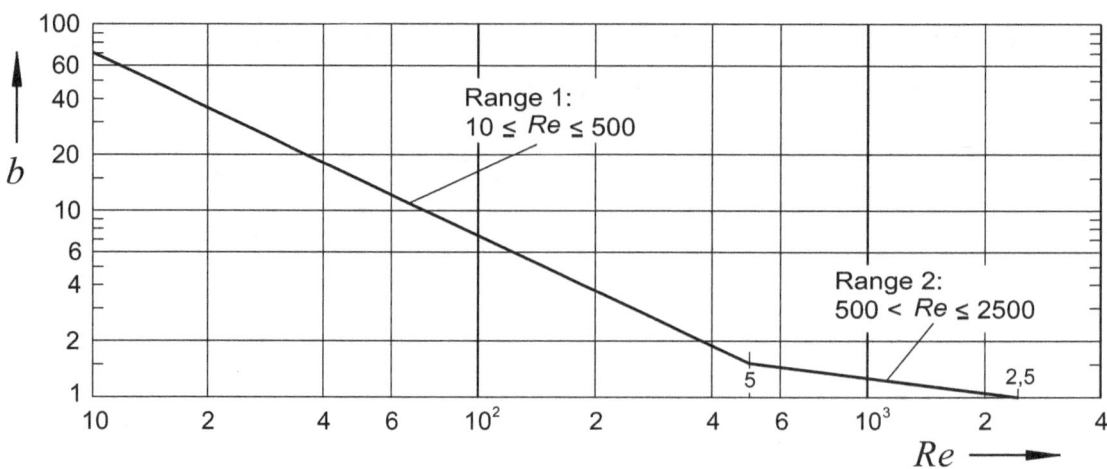

Figure 2.16: Correction factor b as a function of Reynolds-number Re (*Chaimowitsch*)

In Figure 2.16 two *Reynolds*-number ranges are highlighted. For these ranges, the following equations can specify

Range 1: $10 \leq Re \leq 500$ $\qquad b = 668 \cdot Re^{-0.98}$ (2.68)

Range 2: $500 < Re \leq 2500$ $\qquad b = 7,549 \cdot Re^{-0.258}$ (2.69)

Also, the pressure losses occurring in **valves** for hydraulic systems are taken into account in order to determine the pressure losses occur, particularly if the system contains a large number of valves.

The manufactures of valves have **valve characteristics** for each type of valve in form of diagrams. From this, the pressure loss Δp_V can be taken in dependency of the volume flow Q. It should be noted that the valve characteristics are valid for a certain **oil viscosity**.

Figure 2.17 shows as an example valve characteristic of spring-loaded non return valves (company *Mannesmann Rexroth*, nominal size 15, type M-SR and series 1 X). Here the oil viscosity is $v = 41 \, \text{mm}^2/\text{s}$ at $t = 50\,°C$. Different spring rates are marked with the numbers „05" „15" „30".

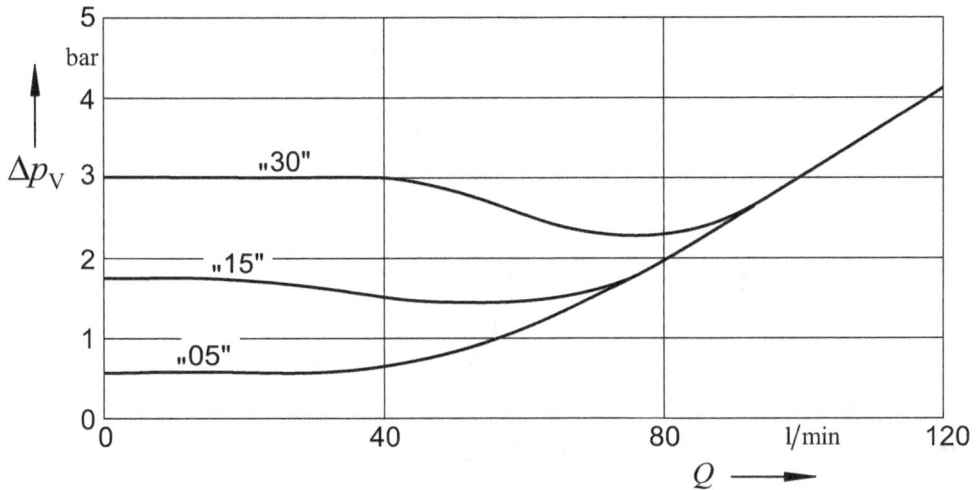

Figure 2.17: Valve characteristics of spring loaded non return valves (*Mannesmann Rexroth*)

If the hydraulic oil has a different cinematic viscosity (v_2) as the one that based on the valve characteristics (v_1), then according to *Bauer* for **laminar flow** is

$$\Delta p_{V2} = \Delta p_{V1} \frac{v_2}{v_1} \qquad (2.70)$$

For **turbulent flow** is

$$\Delta p_{V2} = \Delta p_{V1} \left(\frac{v_2}{v_1}\right)^{0,25} \qquad (2.71)$$

NOTE: More information about resistance coefficients for fittings can be found at *Chaimowitsch, Herning, Panzer/Beitler* and *Wärmetechnische Arbeitsmappe*. Valve characteristics are preferably provided in vendor-specific documentations.

A piping system generally includes pipes, fittings and valves. For such a system can be written in general form

$$p_1 - p_2 = \frac{\rho}{2}\left(v_2^2 - v_1^2\right) + \rho \cdot g \left(z_2 - z_1\right) + \ldots$$

$$\cdots \sum_i \Delta p_{Ri} + \sum_i \Delta p_{Fi} + \sum_i \Delta p_{Vi} \qquad (2.72)$$

Thus, the pressure difference $p_1 - p_2$ between two flow cross-sections - including pipes, fittings and valves - can be calculated.

2.12 Flows through Throttling Devices - Flow Measurement

To determine the volume flow Q throttling devices can be used. These will be installed at a suitable position in the piping system. Throttling devices are **orifice**, **nozzle** or **venturi-tube**, shown in Figure 2.18.

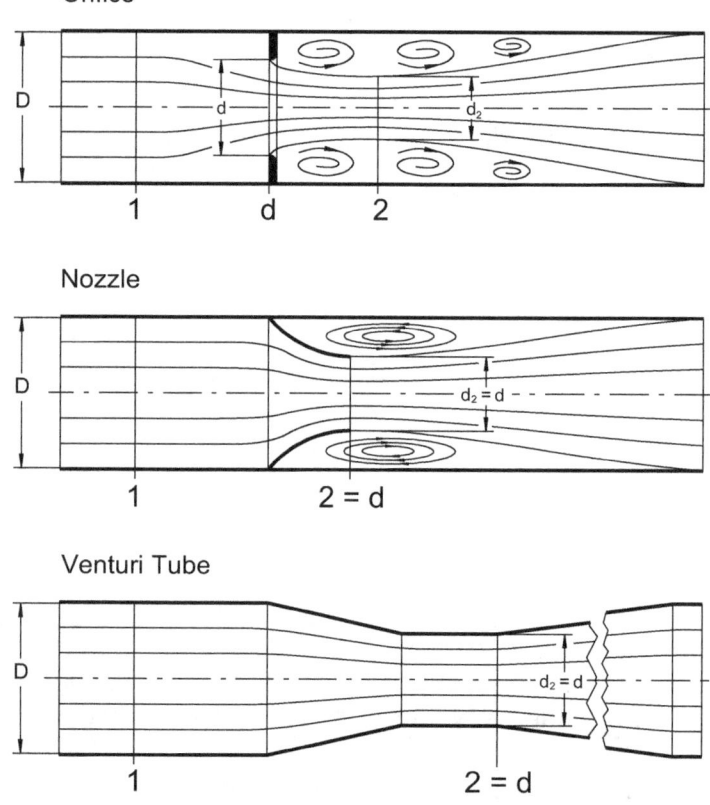

Figure 2.18: Orifice, nozzle and venturi-tube

Physical Principles

The effect of these three devices is based on **Bernoulli-equation** and **continuity equation**. When using the *Bernoulli*-equation in the pressure-form we obtain for the cross-sectional areas 1 and 2

$$p_1 + \rho \cdot g \cdot z_1 + \rho \frac{v_1^2}{2} = p_2 + \rho \cdot g \cdot z_2 + \rho \frac{v_2^2}{2} \qquad (2.73)$$

Solving this equation for v_2 (taking into account $z_1 = z_2$), we obtain

$$v_2 = \frac{1}{\sqrt{1 - \left(\frac{v_1}{v_2}\right)^2}} \sqrt{\frac{2(p_1 - p_2)}{\rho}} \qquad (2.74)$$

According to the continuity equation (2.41) we have in the present case $v_1/v_2 = A_2/A_1$. That gives

$$v_2 = \frac{1}{\sqrt{1 - \left(\frac{A_2}{A_1}\right)^2}} \sqrt{\frac{2(p_1 - p_2)}{\rho}} \qquad (2.75)$$

Thus we obtain for **frictionless** flow

$$Q = A_2 \cdot v_2 = \frac{A_2}{\sqrt{1 - \left(\frac{A_2}{A_1}\right)^2}} \sqrt{\frac{2(p_1 - p_2)}{\rho}}$$

$$= \frac{d_2^2 \frac{\pi}{4}}{\sqrt{1 - \left(\frac{d_2}{D}\right)^4}} \sqrt{\frac{2(p_1 - p_2)}{\rho}} \qquad (2.76)$$

Due to the non-consideration of the influence of friction, this equation cannot be used in practice. Based on equation (2.76) must be used in order to receive realistic results in accordance with **DIN EN ISO 5167-1** *Through-flow measurement of fluids with throttle devices* for determining the volume flow the following equation (friction effects are taken into account)

$$\boxed{Q = C \frac{d^2 \frac{\pi}{4}}{\sqrt{1-\left(\frac{d}{D}\right)^4}} \sqrt{\frac{2 \cdot \Delta p}{\rho}}} \quad (2.77)$$

In equation (2.77) C is called **flow coefficient**, d is the inner diameter of the throttle cross-section and D the upstream inner diameter of the tube (in case of venturi tube D is the diameter of the inlet cylinder). d, D are the diameters under operating conditions (if high temperatures are present the diameters will increase in comparison with the room temperature), β is the ratio of these two diameters.

The ratio $\boxed{\dfrac{C}{\sqrt{1-\left(\dfrac{d}{D}\right)^4}} = \dfrac{C}{\sqrt{1-\beta^4}}}$ is called **flow rate factor**.

The pressure difference Δp in equation (2.77) is measured as a differential pressure. This means that the pressures prevailing at the measurement points of the throttle device are not measured individually: the pressure difference is detected by measuring the difference between them.

The **flow coefficient** C is calculated using the so-called *Reader-Harris/Gallagher*-Equation in **DIN EN ISO 5167-1**. This standard, which specifies the particular geometry of the throttle devices and the exact positions of the pressure measuring points, is to use.

Physical Principles

2.13 Gap Flows

Components of hydraulic systems (for example directional valves and flow control valves, require for their proper function gaps at certain points. These are usually arranged because of their throttling effect between spaces with different pressure levels.

Equations for the calculation of velocity, volume flow and power loss are subsequently presented for two types of gap flows. The equations are based on the conditions **laminar** and **isothermal gap flow** at relatively wide gaps with small gap heights ($2\,\mu m \leq h \leq 20\,\mu m$). The laminar flow assumption is justified because the Reynolds numbers are far due to the small gap height below the critical Reynolds numbers.

In Figure 2.19, the gap is formed by two parallel plates, wherein both plates are stationary.

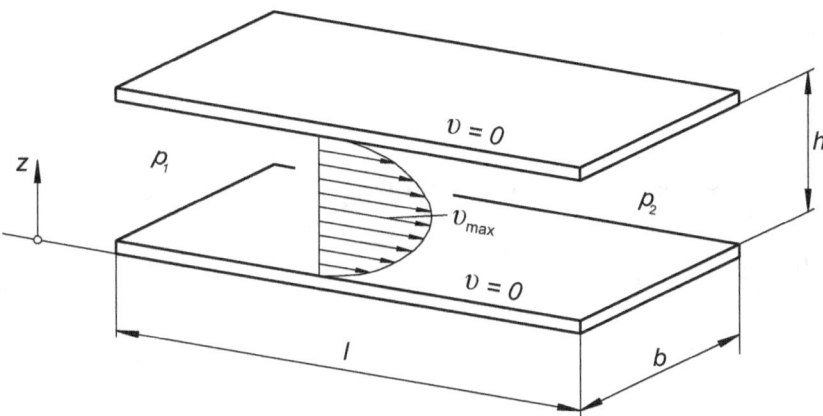

Figure 2.19: Velocity distribution over the gap height (laminar flow)

The gap flow caused by means of the pressure difference $\Delta p = p_1 - p_2$ with $p_1 > p_2$. We speak in this case of a gap flow through a pressure gradient that generates a **leakage flow**.

Over the gap height h is present a **parabolic velocity distribution** which can be expressed as a function of coordinate z (*Ivantysyn*)

$$v(z) = -\frac{\Delta p}{2 \cdot \eta \cdot l}\left(z^2 - h \cdot z\right) \qquad (2.78)$$

The maximum value of the velocity is in the middle of the gap at $z = h/2$

$$v(z = h/2) = v_{max} = \frac{1}{8}\frac{\Delta p \cdot h^2}{\eta \cdot l} \qquad (2.79)$$

The equation for the volume flow through the gap is obtained from

$$Q = \int_{z=0}^{z=h} b \cdot v(z)\, dz \qquad (2.80)$$

The solution of the integral is called the **gap formula**

$$Q = \frac{1}{12}\frac{\Delta p \cdot b \cdot h^3}{\eta \cdot l} \qquad (2.81)$$

The **average velocity** in the gap is calculated using

$$v_m = \frac{Q}{b \cdot h} = \frac{1}{12}\frac{\Delta p \cdot h^2}{\eta \cdot l} = \frac{2}{3}v_{max} \qquad (2.82)$$

The **power loss** in the gap is

$$P_V = Q \cdot \Delta p = \frac{1}{12}\frac{\Delta p^2 \cdot b \cdot h^3}{\eta \cdot l} \qquad (2.83)$$

The gap formula of equation (2.81) is valid under the assumption that the gap width b is large in relation to the gap height h. In this case, the influences of the lateral boundary walls can be neglected. The calculated volume flow by using the gap formula is to be multiplied by a **correction value** k if the b/h-values are small.

Physical Principles

$$Q = \frac{1}{12} \frac{\Delta p \cdot b \cdot h^3}{\eta \cdot l} k \qquad (2.84)$$

Thoma recommends the following values for k

$b/h = 10$: $k = 0{,}94$ $\qquad$ $b/h = 5$: $k = 0{,}88$
$b/h = 3$: $k = 0{,}79$ $\qquad$ $b/h = 2$: $k = 0{,}69$
$b/h = 1$: $k = 0{,}42$

The volume flow through the gap increases with the third power of the gap height h. This is shown by the equations (2.81) and (2.84). Small gap height changes during operation of hydraulic components (for instance due to different thermal expansion coefficients of the materials involved) therefore take great impact on the gap flow.

In Figure 2.20 the gap is also formed by two parallel plates. Now, however, rests only one of the plates; the other is moved with the velocity v_{Plate}. A **pressure difference** does not exist $(p_1 = p_2)$. This is a gap flow caused by a moving plate; it will be a **drag stream** generated.

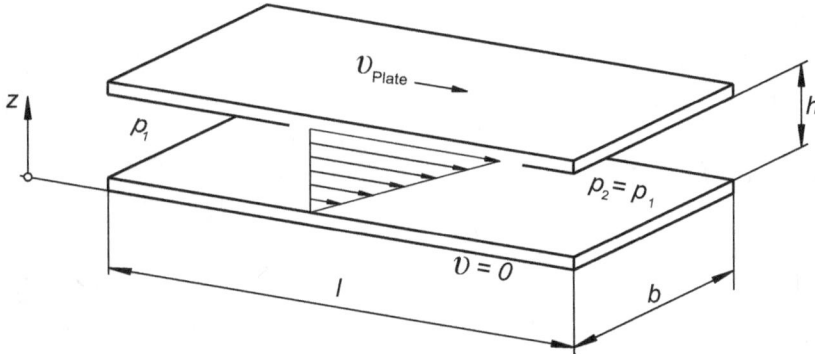

Figure 2.20: Velocity distribution over the gap height (laminar flow)

Over the gap height h there is a linear velocity distribution. This can be expressed as a function of the coordinate z

$$v(z) = v_{\text{Plate}} \frac{z}{h} \qquad (2.85)$$

The maximum value of the velocity is

$$v(z=h) = v_{\max} \qquad (2.86)$$

This is due to the no-slip condition equal to the velocity of the moving plate ($v_{\max} = v_{\text{Plate}}$). The **drag stream** is calculated by

$$Q = \frac{1}{2} b \cdot h \cdot v_{\text{Plate}} \qquad (2.87)$$

The following applies to the **average velocity** in the gap

$$v_m = \frac{Q}{b \cdot h} = \frac{1}{2} v_{\text{Plate}} \qquad (2.88)$$

The **power loss** in the gap is

$$P_V = F \cdot v_{\text{Plate}} = \eta \frac{v_{\text{Plate}}^2}{h} b \cdot l$$

(2.89)

with the **drag force** (see equation (2.52)).

$$F = \tau \cdot b \cdot l = \eta \frac{v_{\text{Plate}}}{h} b \cdot l \qquad (2.90)$$

The friction caused by the gap losses are converted into heat. The increase in temperature reduces the viscosity of the oil, so that the above equations underlying assumption of isotherm flow does not correspond to reality. The equations are to be considered from this point as approximate equations.

Physical Principles

Figure 2.21 shows qualitatively the velocity profiles in the gap between two parallel plates at pressure difference $\Delta p = p_1 - p_2$ ($p_1 > p_2$); the upper plate moves with the velocity v_{Plate}.

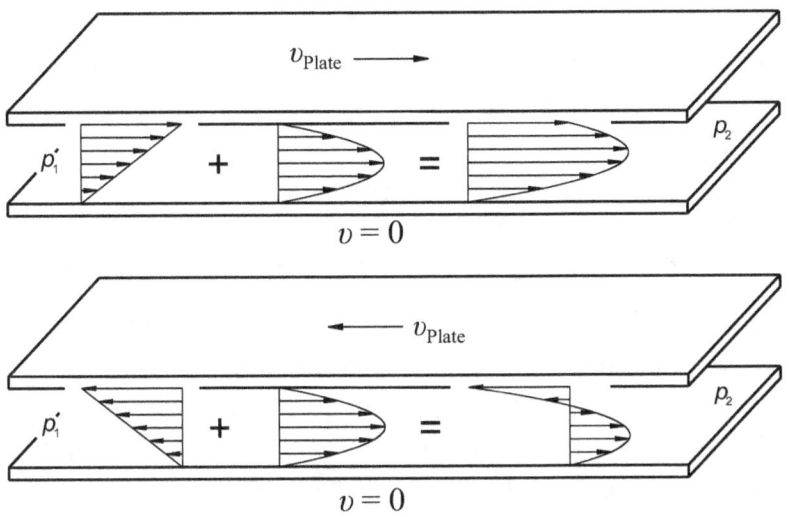

Figure 2.21: Gap flow between two parallel plates at differential pressure and movement of the upper plate (laminar flow)

Figure 2.21 (top) shows the case of moving the top plate to the right with the velocity v_{Plate} at a standstill lower plate. The linear profile of the drag flow is superimposed the parabolic profile caused by the pressure difference $\Delta p = p_1 - p_2$ ($p_1 > p_2$). The resulting flow profile is formed from the adding together the individual profiles. It can be seen that the maximum value of the flow velocity has shifted to the top plate.

Figure 2.21 (below) shows the case of moving the top plate to the left with the velocity v_{Plate} at a standstill lower plate. Here, too, the linear profile of the drag flow is superimposed the parabolic profile caused by the pressure difference Δp. Due to the different velocity vectors directed results a resulting velocity profile, from which it is seen that flow particles are moving in the vicinity of the upper plate to the left and in the vicinity of the bottom plate to the right. On seals directed flows, caused by the effects described with Figure 2.21, may lead to high stresses and can damage the seals.

> NOTE: To calculate the flow in the gap (especially for circular gaps) are more equations to find by *Ivantysyn*.

2.14 Hydraulic Resistances

The **hydraulic resistance** of a component used in a hydraulic system is defined in analogy to the electrical resistance $R = U/I$ as

$$R = \Delta p / Q \qquad (2.91)$$

In this equation Δp is the component (e.g. valve) applied pressure difference (pressure drop) and Q is the volume flow flowing through the component. The hydraulic resistance of a component depends on many influencing factors. In particular, take the **form of flow** (laminar or turbulent), **viscosity** and **temperature** of the oil influence on the hydraulic resistance. Figure 2.22 shows components with their hydraulic resistances in **series connection** and **parallel connection**.

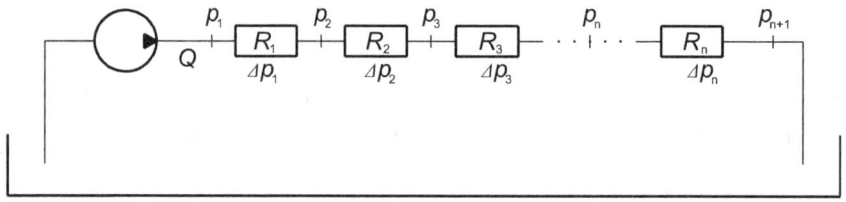

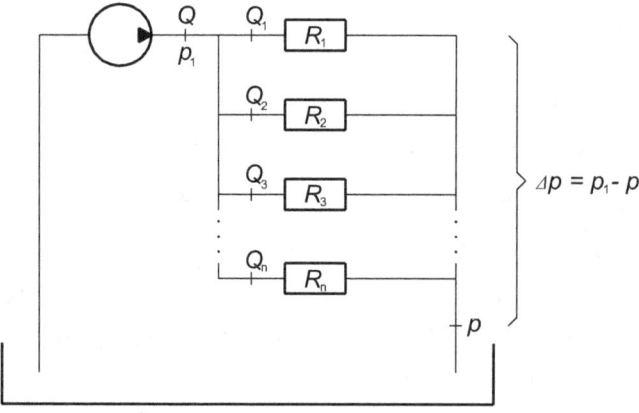

Figure 2.22: Hydraulic components in series (top) and in parallel connection

Physical Principles **45**

In **series connection** in Figure 2.22 (top) flows through all components the same volume flow Q. The prevailing pressure before each component can be expressed by the pressure behind the component added to the pressure loss:

$$p_1 = p_2 + \Delta p_1 \quad p_2 = p_3 + \Delta p_2 \quad p_3 = p_n + \Delta p_3 \quad p_n = p_{n+1} + \Delta p_n \tag{2.92}$$

The addition of the left and right sides of equations (2.92) results in

$$p_1 + p_2 + p_3 + \ldots + p_n =$$
$$p_2 + \Delta p_1 + p_3 + \Delta p_2 + p_n + \Delta p_3 + \ldots + p_{n+1} + \Delta p_n \tag{2.93}$$

It follows

$$p_1 - p_{n+1} = \Delta p_1 + \Delta p_2 + \Delta p_3 + \ldots + \Delta p_n = \sum_{i=1}^{n} \Delta p_i \tag{2.94}$$

The **pressure loss** in the series connection is equal to the sum of the single pressure losses. From equation (2.94) is follows with $\Delta p_i = Q \cdot R_i$

$$p_1 - p_{n+1} = \sum_{i=1}^{n} Q \cdot R_i = Q \sum_{i=1}^{n} R_i = Q \cdot R_{tot} \tag{2.95}$$

Then the **total resistance** can be expressed in series connection by means of

$$R_{tot} = \sum_{i=1}^{n} R_i = \frac{p_1 - p_{n+1}}{Q} \tag{2.96}$$

When arranging components in parallel connection according to Figure 2.22 (below) to each component, the same pressure differential $\Delta p = p_1 - p$ exists.

Further, the sum of the component flow rates is equal to the **total flow** provided by the hydraulic pump

$$Q_1 + Q_2 + Q_3 + \ldots + Q_n = Q \tag{2.97}$$

Using the relations $Q_1 = \Delta p/R_1 \quad Q_2 = \Delta p/R_2 \quad Q_3 = \Delta p/R_3$ we get

$$\frac{\Delta p}{R_1} + \frac{\Delta p}{R_2} + \frac{\Delta p}{R_3} + \ldots + \frac{\Delta p}{R_n} = Q \tag{2.98}$$

It is with $Q = \Delta p/R_{tot}$

$$\frac{\Delta p}{R_1} + \frac{\Delta p}{R_2} + \frac{\Delta p}{R_3} + \ldots + \frac{\Delta p}{R_n} = \frac{\Delta p}{R_{tot}} \tag{2.99}$$

The **total hydraulic resistance** in a parallel connection can thus be found from the following relationship

$$\frac{1}{R_{tot}} = \frac{1}{R_1} + \frac{1}{R_2} + \frac{1}{R_3} + \ldots + \frac{1}{R_n} \tag{2.100}$$

Are two components arranged in parallel connection then applies

$$Q_1 = Q \frac{R_2}{R_1 + R_2} \qquad Q_2 = Q \frac{R_1}{R_1 + R_2} \tag{2.101}$$

Are three components arranged in parallel then applies

$$Q_1 = Q \frac{R_2 \cdot R_3}{R_1 \cdot R_2 + R_2 \cdot R_3 + R_1 \cdot R_3} \tag{2.102}$$

$$Q_2 = Q \frac{R_1 \cdot R_3}{R_1 \cdot R_2 + R_2 \cdot R_3 + R_1 \cdot R_3} \tag{2.103}$$

$$Q_3 = Q \frac{R_1 \cdot R_2}{R_1 \cdot R_2 + R_2 \cdot R_3 + R_1 \cdot R_3} \tag{2.104}$$

Physical Principles 47

2.15 Compressibility and Compression Module

To explain the term **compressibility** Figure 2.23 is used. It shows a hydraulic cylinder with oil-filled piston-side cylinder chamber. There are made the following assumptions: All parts of the cylinder are rigid, the piston is sealed leak-free and guided without friction. The outflow of the oil from the cylinder chamber is prevented by the closed valve on the left.

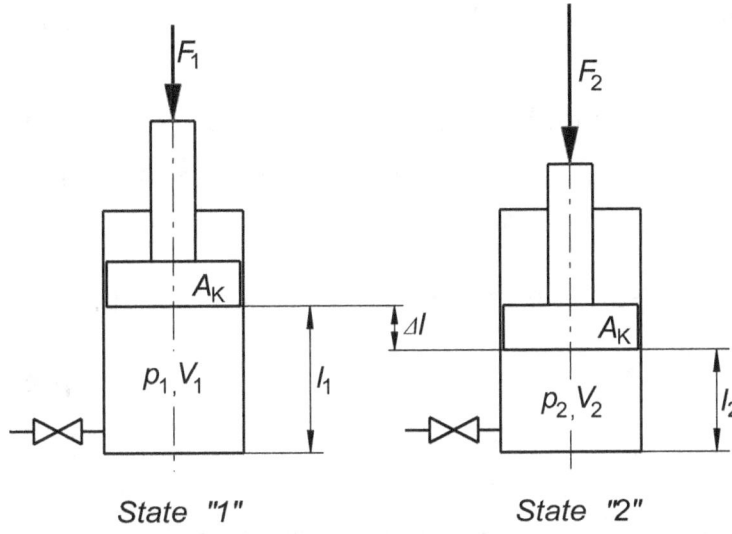

Figure 2.23: Volume change due to the influence of the compressibility

In Figure 2.23 (left) is the oil under the pressure of $p_1 = F_1/A_K$ (F_1 is the force acting on the oil via the piston area A_K). In the state "1" the oil has the volume V_1.

If the force F_1 will be increased by ΔF the oil reached the state "2". The force on the piston is now $F_2 = F_1 + \Delta F$. This is shown in Figure 2.23 (right). In the state "2" the oil has the volume V_2. Due to the **compressibility**, which is a physical property of the oil, we note a decrease of the volume. The result is

$$V_2 < V_1 \tag{2.105}$$

$$V_2 - V_1 = \Delta V_F = (l_2 - l_1) A_K = -(l_1 - l_2) A_K = -\Delta l \cdot A_K < 0 \tag{2.106}$$

At state "2" the oil has a smaller volume than in state "1". The volume change $\Delta V_F < 0$ is called **volume of compression**; $\Delta l = l_1 - l_2$ is the way the piston moves down.

In general, assuming $T = \text{const.}$ (**isothermal** change of state) for the total change in volume of a liquid applies

$$dV = -V \cdot \beta_p \cdot dp = -\frac{V}{K} dp \qquad (2.107)$$

Designations: β_p is the **isothermal compressibility coefficient** and $K = 1/\beta_p$ is the **compression module** of the liquid, also known as the **true compression module**.

According to equation (2.107) results for the true compression module

$$K = -V \frac{dp}{dV} = -\frac{V}{\dfrac{dV}{dp}} \qquad (2.108)$$

Herein dV/dp is a measure of the gradient of the tangent to the volume-pressure curve. Equation (2.108) can therefore be expressed as follows

$$K = -\frac{V}{\tan\alpha \cdot m_S} \qquad (2.109)$$

Herein m_S is the scale factor of the diagram (explanation below, see Figure 2.24). In the hydraulics the true compression module will rarely use (for example in vibration analysis of oil columns). Must find consideration the compressibility of the hydraulic oil, is usually sufficient to use the **average compression module**, also known as **secant compression module**. Taking into account equations (2.108) and (2.109) we will find for the average compression module the relationship

Physical Principles

$$K_S = -\frac{V_1}{\frac{V_2 - V_1}{p_2 - p_1}} = -\frac{V_1}{\tan \alpha_S \cdot m_S} \qquad (2.110)$$

The diagram in Figure 2.24 is intended to explain the determination of the **true** and the **average compression module**. It shows the **pressure-volume-curve** of a special sort of hydraulic oil for the pressure range 0 bar $< p_e <$ 1000 bar (overpressure) at $t = 10°C$. At atmospheric pressure ($p_e = 0$ bar), there is present an oil volume of $V = 1\,m^3$ (initial volume).

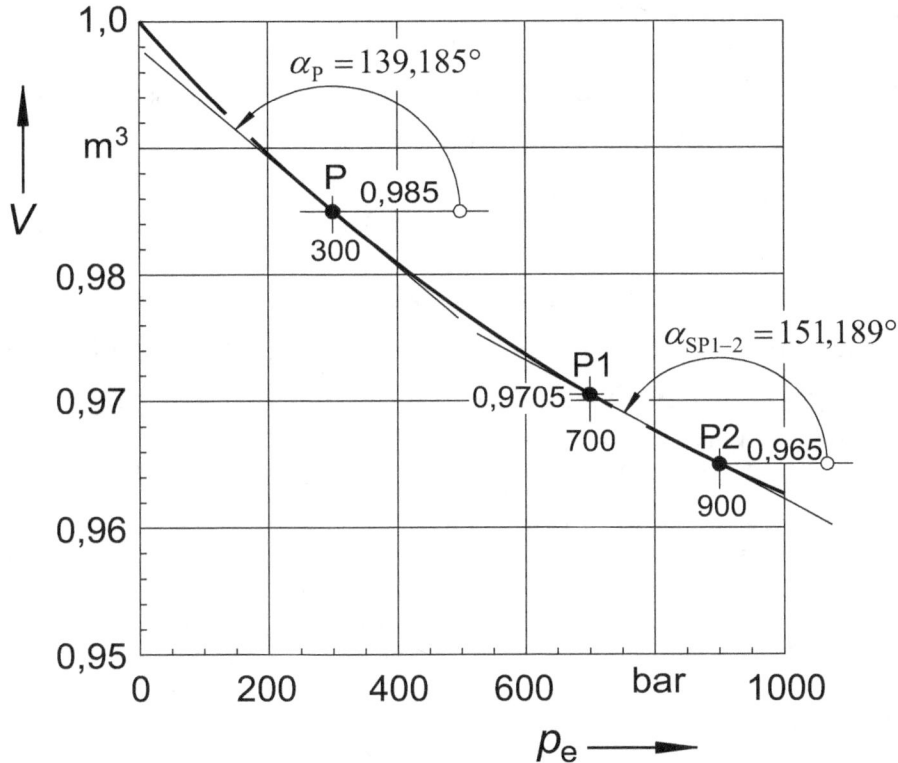

Figure 2.24: Volume-pressure curve of a special sort of hydraulic oil

If, for example the oil pressure is $p_{eP} = 300$ bar, the oil volume is $V_P = 0,9850\,m^3$ (point P). The tangent at point P is inclined to the abscissa axis at the angle $\alpha_p = 139,185°$. Equation (2.109) gives thus for the point P for the true compression module taking into account the scale factors of the diagram

$$K_P = -\frac{V_P}{\tan \alpha_P \cdot m_S} = -\frac{0,9850 \text{ m}^3}{\tan 139,185° \dfrac{0,05 \text{ m}^3/75 \text{ mm}}{1000 \text{ bar}/75 \text{ mm}}} = 22811 \text{ bar} \qquad (2.111)$$

As the example shows, the determination of the true compression module requires the definition of a point on the volume-pressure curve. The present values for V_P and α_P then determine (in connection with the diagram scale factor m_S) the valid value only for this point.

For the determination of an average compression module a pressure range must be defined. For example: If the pressures are $p_{eP1} = 700 \text{ bar}$ and $p_{eP2} = 900 \text{ bar}$ a pressure range is defined (Figure 2.24); the oil volumes for these pressures are $V_{P1} = 0,9705 \text{ m}^3$ and $V_{P2} = 0,9650 \text{ m}^3$ (see points P1 and P2). A secant (defined by the points P1 and P2) is inclined to the abscissa axis at the angle $\alpha_S = 151,189°$. According to equation (2.110) the result for the average compression module (taking into account the scale factor m_S of the diagram) is

$$K_{S\,P1\text{-}P2} = -\frac{V_{P1}}{\tan \alpha_{S\,P1\text{-}P2} \cdot m_S} = -\frac{0,9705 \text{ m}^3}{\tan 151,189° \dfrac{0,05 \text{ m}^3/75 \text{ mm}}{1000 \text{ bar}/75 \text{ mm}}} \qquad (2.112)$$

$K_{S\,P1\text{-}P2} = 32291 \text{ bar}$

As the example shows the calculation of the average compression module must be preceded the definition of a pressure range. The numerical value found for the average compression module is then valid for the specified pressure range.

Solving equation (2.110) for V_2 gives

Physical Principles 51

$$V_2 = V_1 \left(1 - \frac{p_2 - p_1}{K_S}\right) \qquad (2.113)$$

Knowing K_S V_1 p_1 and p_2 we are able to calculate the volume V_2 at the pressure p_2. If, in equation (2.113), the numerical values of the above example used we obtain

$$V_2 = 0{,}9705 \text{ m}^3 \left(1 - \frac{(900 - 700) \text{ bar}}{35291 \text{ bar}}\right) = 0{,}965 \text{ m}^3 \; [= V_{P2}] \qquad (2.114)$$

2.16 Cavitation

There are two different types of cavitation: the air bubble and the vapor bubble cavitation. Both types of cavitation have similar negative effects on components of hydraulic systems.

Air bubble cavitation: Fluids have the property of absorbing gases in it. One speaks in this connection of the gas absorption capacity of the liquids. Hydraulic oils absorb in particular air. In addition to the dissolved form, the air can also occur in the form of air bubbles in the oil. This happens when local the static pressure of the oil does drop to the gas release pressure (saturation pressure). Then the capacity of the oil for air is exhausted. Pressure reductions of the oil can occur at constrictions of hydraulic components due to increased flow velocities present there (e.g. in valves and hydraulic pumps). Does it come after the constriction due to the expansion of the flow cross-section because of the reduction of the flow velocity to a rise in the pressure then the bubbles will collapse abruptly in form of an implosion.

Vapor bubble cavitation: Is used in technology the term cavitation, so that the vapor bubble cavitation is usually meant. This occurs when vapor bubbles are formed through a reduction in the static pressure up to or below the vapor pressure. Here too the drop in pressure is affected by the increased flow velocities existing at constrictions in hydraulic components. The after the constriction decreasing flow velocity lets the pressure rise again, so that the vapor bubbles (like the bubbles in the air bubble cavitation) will collapse abruptly in form of an implosion.

So cavitation means: **Formation of bubbles** (air or vapor bubbles) at constrictions in hydraulic components caused by pressure reduction and by the sudden collapse of the bubbles after leaving the constriction by re-increases the pressure. Figure 2.25 is intended to illustrate the occurrence of cavitation in principle in a pipe with constriction.

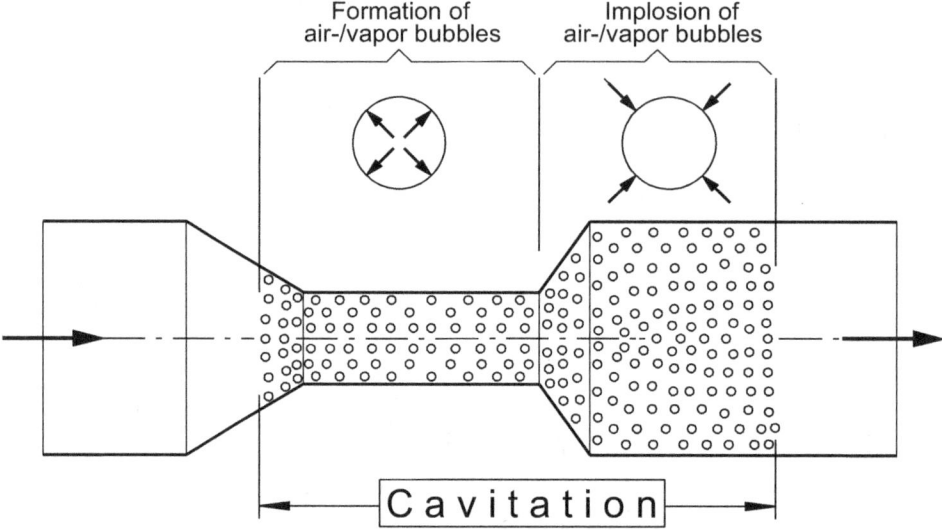

Figure 2.25: Occurrence of cavitation in a pipe with constriction (in principle)

Whether it is an air bubble cavitation or a vapor bubble cavitation depends on whether rather gas release pressure or rather the vapor pressure at the constriction of the hydraulic component is achieved.

If cavitation is present in hydraulic components, occur because of the sudden collapse of the bubbles pattering noise (cavitation noise) and vibration. Due to the sudden decrease in volume of the bubbles microscopic liquid jets are produced in high-frequency sequence. These cause upon impact with walls locally extremely high pressures. As a result, the material is eroded. This process is known as cavitation erosion, which is the main cause of defects in material damage according to current knowledge.

Physical Principles **53**

Cavitation can also lead to the reduction in performance of hydraulic components. It can, for example, in hydraulic pumps come due to cavitation to the reduction of flow cross sections, which leads to the change of the pump characteristics. With the implosion of the bubbles, a local increase in temperature is associated, which may be so high that it may lead to spontaneous ignition of the oil. The local increase in temperature as a result of cavitation possibly changes the properties of the hydraulic oil (aging caused by cavitation).

Cavitation should be avoided in components of hydraulic systems because of the negative impacts. This can be done by suitably selecting of the components and flow calculations using suitable computer programs.

3 Basic Structure of a Hydraulic System

By Figure 3.1 will be explained below the basic structure of a hydraulic system.

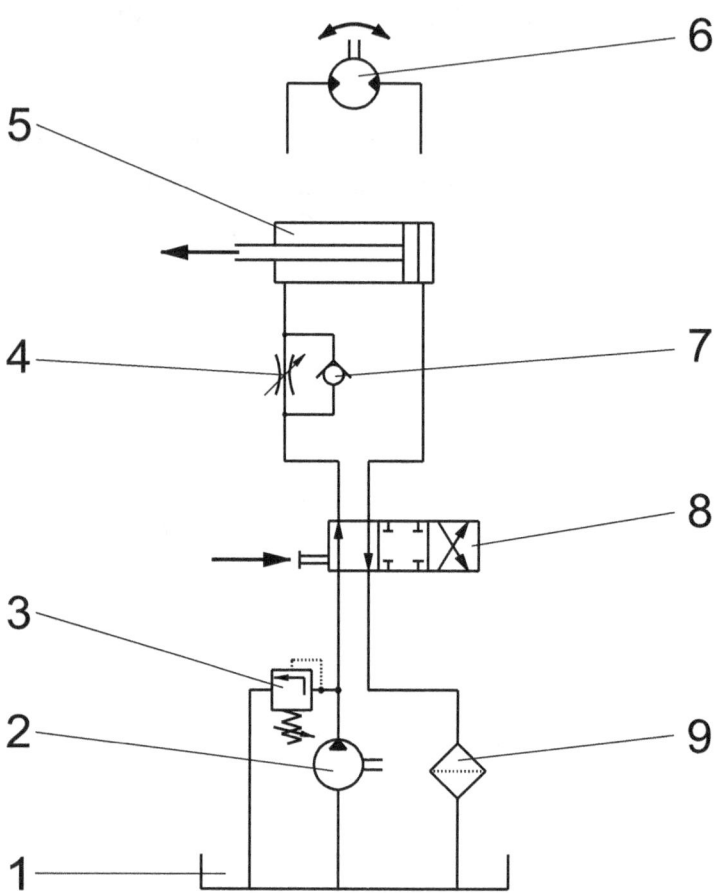

Figure 3.1: Basic Structure of a Hydraulic System

Figure 3.1 is made using the standardized symbols according **DIN ISO 1219-1**.

Basic Structure of a Hydraulic System

From the **tank** (1) the **hydraulic pump** (2) sucks oil on. Via the drive shaft of the pump mechanical energy is fed to. This is for the most part converted into hydraulic energy which is in the oil. At the outlet port of the pump the hydraulic energy is mainly present as pressure energy in the oil flow. The oil flow passes through the **directional control valve** (8) and the **flow control valve** (4) on the piston side of the **hydraulic cylinder** (5); the directional control valve is in the position "right": the piston rod moves out and the oil of the piston rod side is conducted via the directional control valve and the **filter** (9) to the tank. In this operation, the **check valve** (7) is closed; on it, the pressure generated by the hydraulic pump, acts.

To retract the piston rod, the directional control valve is switched into the position "left". The oil flow coming from the hydraulic pump now enters on the piston rod side of the hydraulic cylinder: the piston rod retracts. The oil which is located on the piston rod side in the hydraulic cylinder is supplied to the now open non-return valve to the tank.

The task of the directional control valve, therefore, is to control the path of the flow to allow the required extension or retraction of the piston rod of the hydraulic cylinder.

The oil pressure generated by the pump depends on the load acting on the piston rod. In this context one speaks of the load resistance of the oil consumer. The oil consumer is in this case the hydraulic cylinder.

The **pressure relief valve** (3) is adjusted to the maximum allowed pressure of the hydraulic system. If the adjusted pressure is reached, it opens and a part of the oil flow passes back to the tank. A further increase in pressure at the outlet of the pump is thereby prevented. The pressure relief valve thus acts as a safety valve of the hydraulic system.

With the flow control valve can be controlled the flow of oil that enters the hydraulic cylinder. This always takes place in connection with the pressure relief valve. When reducing of the flow rate entering the hydraulic cylinder the velocity of the piston rod is reduced. Is reduced using the adjusting screw of flow control valve the throttle cross section, the pressure between the pressure port of the pump and the pressure relief valve increases. If this pressure reaches the set pressure of the pressure relief valve it will open and only a portion of the pump oil flow reaches the hydraulic cylinder; the other part flows through the pressure relief valve to the tank. The result is a reduction of the extending velocity of the piston rod, which is dependent on the piston area and the entering flow rate of the hydraulic cylinder.

In Figure 3.1 it can be seen above the hydraulic cylinder as an alternative pressure oil consumer a **hydraulic motor** (6) which then is used when no translational movement but a rotational movement is required.

The here presented basic structure of a hydraulic system includes components to the field of oil processing, energy conversion an energy control which can be found in all hydraulic systems.

4 Circuit Diagrams

Here are presented some circuit diagrams of **simple hydraulic systems**.

NOTE: Understanding the circuit diagrams requires knowledge about the meaning of the symbols used according to **DIN ISO 1219-1**.

The hydraulic cylinder seen in Figure 4.1 is controlled by means of a **3/2-directional control valve** which is actuated by hand.

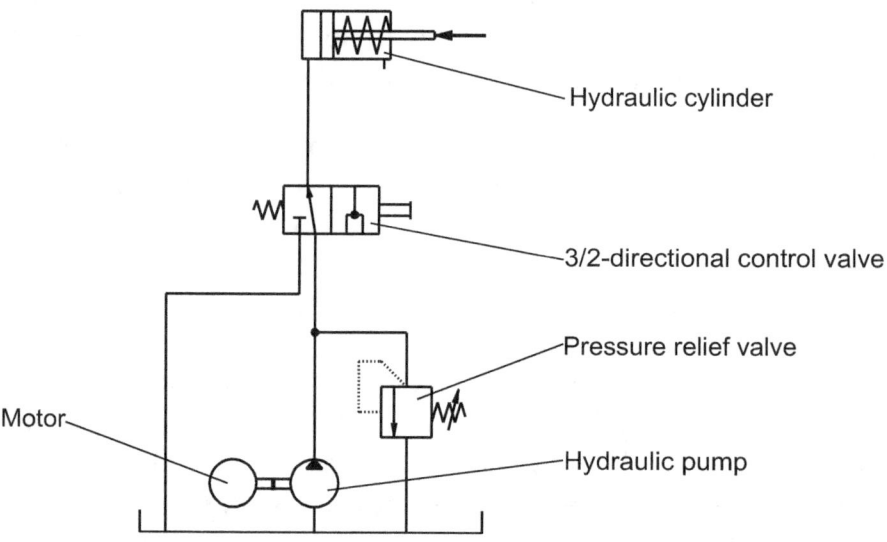

Figure 4.1: Circuit Diagram for controlling a hydraulic cylinder by means of a 3/2-directional control valve

Circuit Diagrams **57**

The 3/2-directional control valve has three ports and two switching positions. The identification of directional control valves is effected by the numbers of ports (in this case: three) and the numbers of switching positions (in this case: two).

Figure 4.1 shows the directional control valve in the switching position "right". In this position the hydraulic pump delivers oil to the piston side chamber of the hydraulic cylinder: the piston rod moves out against the load. When the piston of the hydraulic cylinder reaches its mechanical end position (stop position), the pressure in the pressure pipe rises to such an extent that the pressure value at which the pressure relief valve is sets, is obtained. Then, the whole oil flow coming from the pump flows via the pressure relief valve to the tank. That means: In the pressure relief valve the energy contained in the oil flow is converted into heat by throttling. The stop position is thus characterized by high energy losses in heating the pressure relief valve.

To retract the piston rod, the directional control valve is switched in the switching position "left". In this position the pump flow flows through the directional control valve immediately back to the tank. During retraction of the piston rod, the oil is pressed out of the piston side chamber of the hydraulic cylinder; it flows via point K to the tank. It is mixed with the oil coming from the pump which immediately flows back to the tank. The oil pressure at the outlet of the pump is in this case determined only by the pressure losses in the directional control valve and the pressure losses occurring in the pipe.

Here a single-acting hydraulic cylinder is used. In the piston rod side chamber of the hydraulic cylinder there is a helical compression spring situated. This effected the retraction of the piston rod.

In Figure 4.2 a single-acting hydraulic cylinder is controlled using a **3/3-directional control valve**. The valve has in the switching positions "right" and "left" the same functions as the one in Figure 4.1 used 3/2-directional control valve.

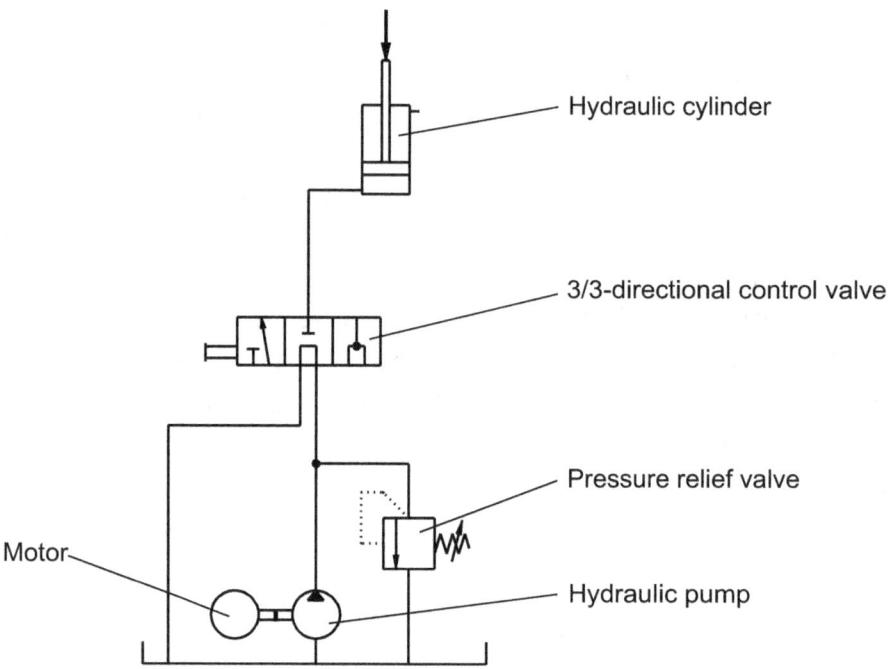

Figure 4.2: Circuit Diagram for controlling a hydraulic cylinder by means of a 3/3-directional control valve

In the switching positions "right" moves out the piston rod under the load. In the **middle switching position** of the 3/3-directional control valve the movement of the piston rod stops due to the flow of the oil to the hydraulic cylinder is interrupted and returned to the tank. Stopping the piston rod is possible any time. For this, the valve must be placed in the middle switching position (holding position). Under load the piston rod during holding up their position may change when the 3/3-directional control valve is not completely leak-free. Pressure losses and heating are low in the holding position. Regarding the switching position "left" reference is made in description for Figure 4.1.

In Figure 4.3 a **synchronizing cylinder** with a hand-operated **4/3-directional control valve** is controlled.

Circuit Diagrams

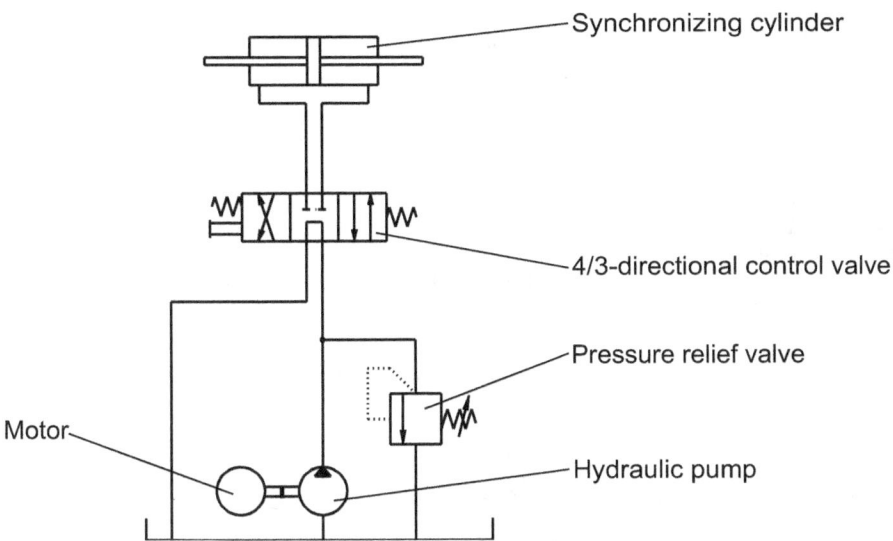

Figure 4.3: Circuit Diagram for controlling a hydraulic cylinder by means of a 4/3-directional control valve

Is the valve in the switching position „right", the right piston rod moves out, the left piston rod retracts. In the switching position „left", the left piston rod moves out, the right piston rod retracts. In the non-actuated position the directional control valve is held by spring force in the central switching position (holding position). This means, that the piston and the piston rods not move. In this position prevents the locking effect of the directional control valve when external forces act on the piston rods whose movement. In the holding position flows the oil from the pump via the directional control valve neatly without losses back to the tank: the oil gets only a low temperature rise.

Also for controlling the synchronizing cylinder in Figure 4.4 a hand-operated spring-centered 4/3-directional control valve is used.

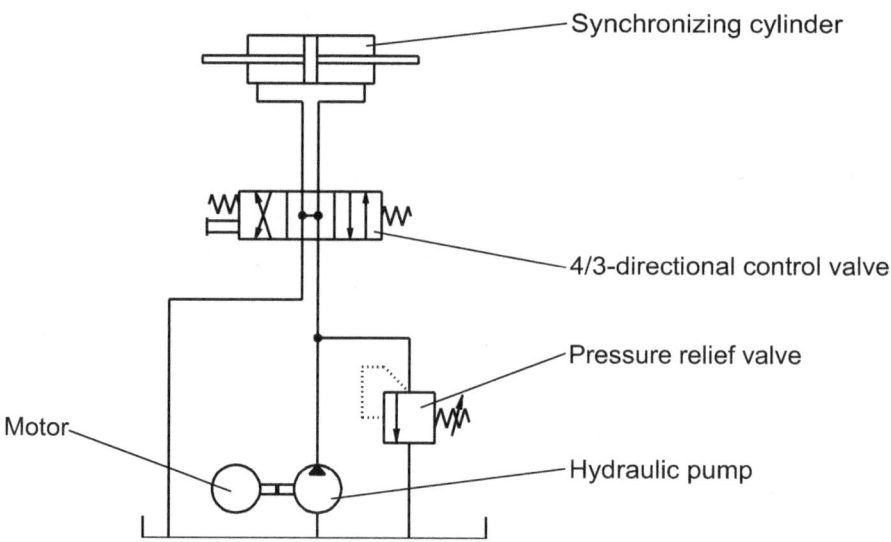

Figure 4.4: Circuit Diagram for controlling a hydraulic cylinder by means of a 4/3-directional control valve

Is the valve in the switching position „left", the left piston rod moves out, the right piston rod retracts. In the switching position „right", the right piston rod moves out, the left piston rod retracts. In the **middle switching position** (the so-called **floating position**) both cylinder chambers and the hydraulic pump are connected to the oil reservoir. In the floating position piston and piston rod can be moved by externally acting forces.

For controlling a **hydraulic motor**, the circuit configuration shown in Figure 4.5 can be used.

Circuit Diagrams

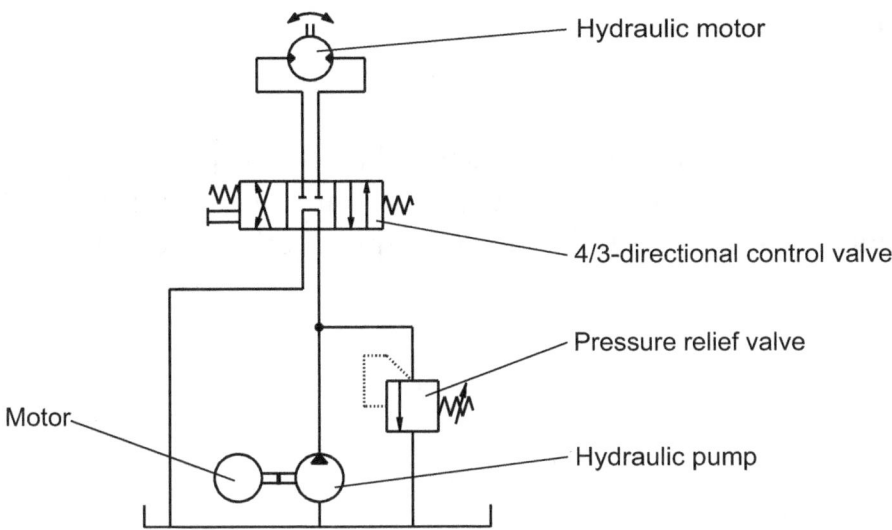

Figure 4.5: Circuit Diagram for controlling a hydraulic motor by means of a 4/3-directional control valve

Here, too, a hand-operated spring-centered **4/3-directional control valve** is used. In the switching positions "right" and "left" oil is fed by the hydraulic pump to the hydraulic motor. The direction of rotation of the hydraulic motor depends on the respective position of the directional control valve. When the directional control valve is in the middle position, the oil flows from the hydraulic pump interrupted. The hydraulic motor then runs down to a standstill. In the middle position, the hydraulic pump delivers the oil flow **nearly lossless** via the directional control valve back to the tank.

5 Examples

Example 1:

The container shown in Figure 5.1 is filled with a liquid. The pistons 1 and 2 are free of friction and leakage. On the upper piston 1 there is a body of mass m. The horizontal mounted piston 2 is acting against a compression spring.

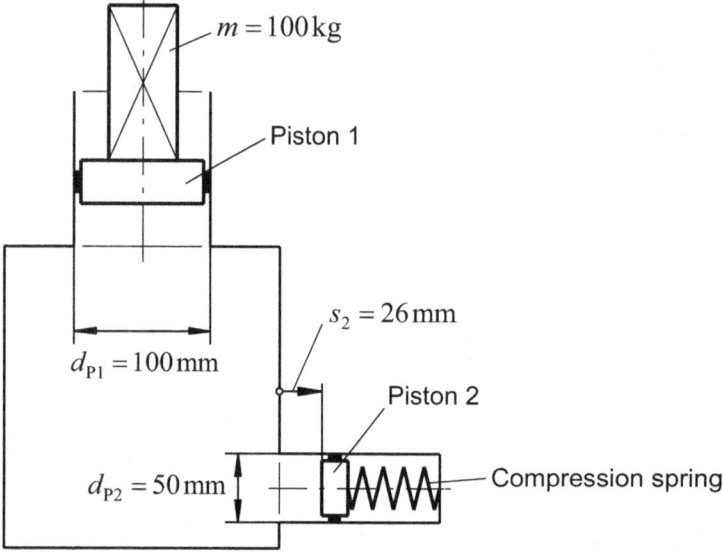

Figure 5.1: Container with two pistons

The data required for the solution are given in Figure 5.1. The liquid in the container is assumed to be incompressible.

Wanted:

The spring rate of the compression spring has to be calculated. The spring is compressed $s_2 = 26\,\text{mm}$ by after placing the body with the mass m on the piston 1.

Solution:

1. In the container-fluid pressure $p_{\text{Flü}}$ prevails. This is determined by the body of mass $m = 100\,\text{kg}$, which acts on the piston 1.

$$p_{\text{Flü}} = \frac{F_{P,1}}{A_{P,1}} = \frac{m \cdot g}{d_{P,1}^2 \frac{\pi}{4}} = \frac{100 \text{ kg} \cdot 9{,}81 \frac{\text{m}}{\text{s}^2}}{0{,}1^2 \text{ m}^2 \frac{\pi}{4}} = 124\,905 \frac{\text{N}}{\text{m}^2}$$

2. This pressure (overpressure) acts according to the law of *Pascal* also on the area of the piston 2 (gravity pressure is disregarded). The piston force is thus

$$F_{P,2} = p_{\text{Flü}} \cdot A_{P,2} = p_{\text{Flü}} \cdot d_{P,2}^2 \frac{\pi}{4} = 124\,905 \frac{\text{N}}{\text{m}^2} 0{,}05^2 \text{ m}^2 \frac{\pi}{4} = 245{,}25 \text{ N}$$

3. The compression spring then has the spring rate

$$R = \frac{F_{P,2}}{s} = \frac{245{,}25 \text{ N}}{26 \text{ mm}} = 9{,}43 \frac{\text{N}}{\text{mm}}$$

Example 2:

A water conducting channel has a lateral drain pipe, which is closed by a circular lid (Figure 5.2).

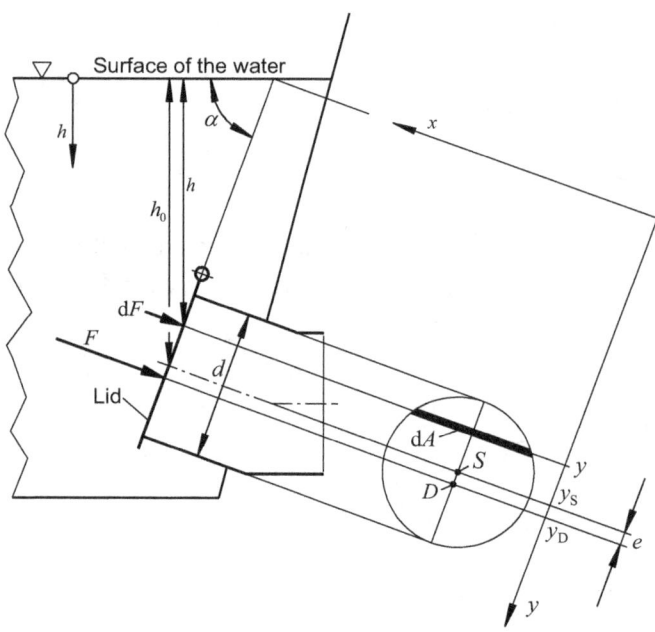

Figure 5.2: Water conducting channel with drain pipe

For more data:

Inner diameter of the drain pipe	$d = 400$ mm
Distance lid center of gravity - surface of the water	$h_S = 2$ m
Density of the water	$\rho = 1000$ kg/m^3
Angle of the lid surface to the water surface	$\alpha = 80°$

Wanted:

The total load F acting on the lid as a result of hydrostatic pressure has to be calculated. Furthermore, the point of attack is to be determined (distance $e = \overline{SD} = y_D - y_S$).

Solution:

1. On the area element dA acts on the side facing the liquid the absolute pressure

$$p_{abs,h} = p_{e,h} + p_{amb}$$

Examples

2. The ambient pressure p_{amb} acts on the back of the lid. Therefore the resultant force on the area element dA is

$$dF = (p_{e,h} + p_{amb})\,dA - p_{amb}\cdot dA = p_{e,h}\cdot dA = \rho\cdot g\cdot h\cdot dA = \rho\cdot g\cdot y\cdot \sin\alpha\cdot dA$$

3. Integration gives

$$F = \rho\cdot g\cdot \sin\alpha \int_A y\,dA$$

4. The integral in this equation represents the static moment (moment of area 1st order) of the pressed area A with respect to the x-axis (see Figure 5.2). It is

$$\int_A y\,dA = y_S \cdot A$$

5. With this the required force is

$$F = \rho\cdot g\cdot \sin\alpha\cdot y_S \cdot A = \rho\cdot g\cdot h_S\,\frac{d^2}{4}\pi$$

$$F = 1000\,\frac{\text{kg}}{\text{m}^3}\,9{,}81\,\frac{\text{m}}{\text{s}^2}\,2\,\text{m}\,\frac{0{,}4^2\,\text{m}^2}{4}\pi = 2\,465{,}5\,\text{N}$$

6. The hydrostatic pressure increases with the depth. The force F therefore does not act in the center of the lid. The point of action (point D) of the force F is shifted by the amount $e = y_D - y_S$ down.

7. The total moment around the x-axis is equal to the sum of the individual moments. It is

$$y_D \cdot F = \int_A y\,dF$$

8. Using the equations found for F and $\mathrm{d}F$ results

$$y_D \cdot \rho \cdot g \cdot \sin\alpha \cdot y_S \cdot A = \int_A y \cdot \rho \cdot g \cdot y \cdot \sin\alpha \, \mathrm{d}A$$

9. It follows

$$y_D = \frac{1}{y_S \cdot A} \int_A y^2 \, \mathrm{d}A$$

10. The integral in this equation is the area moment of 2nd order of the area relative to the x-axis.

With $I_x = \int_A y^2 \, \mathrm{d}A$ we obtain

$$y_D = \frac{I_x}{y_S \cdot A}$$

11. For the calculation of I_x the set of *Steiner* is used. This is

$$I_x = I_{S,x} + y_S^2 \cdot A$$

In this equation $I_{S,x}$ is the area moment of 2nd order with respect to the x-axis lying parallel to the axis of gravity.

12. For y_D thus we obtain

$$y_D = \frac{I_{S,x} + y_S^2 \cdot A}{y_S \cdot A} = \frac{I_{S,x}}{y_S \cdot A} + y_S$$

13. With $I_{S,x} = \pi \frac{d^4}{64}$, $y_S = \frac{h_S}{\sin\alpha}$ and $A = \frac{d^2}{4}\pi$ this equation allows to find the desired distance $e = y_D - y_S$. It is

$$e = \frac{I_{S,x}}{y_S \cdot A} = \frac{\pi \dfrac{d^4}{64}}{\dfrac{h_S}{\sin\alpha} \dfrac{d^2}{4} \pi} = \frac{\sin\alpha}{16} \frac{d^2}{h_S} = \frac{\sin 80°}{16} \frac{400^2 \text{ mm}^2}{2000 \text{ mm}} = 4{,}92 \text{ mm}$$

Example 3:

A pump delivers water from a dam into a higher lying container (Figure 5.3).

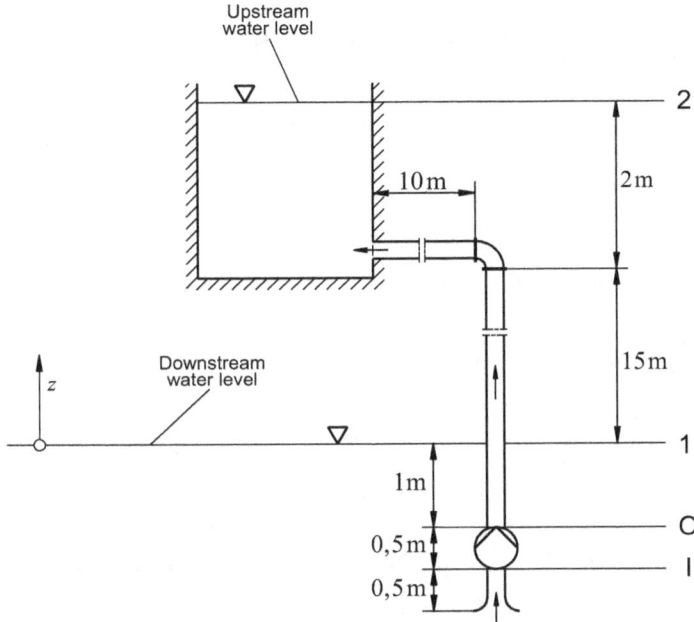

Figure 5.3: Pump delivers water from a dam into a container

For more data:

Internal pipe diameter (suction- and pressure side)	$d = 100$ mm
Density of water	$\rho = 1000 \text{ kg/m}^3$
Atmospheric pressure	$p_{amb} = 1 \text{ bar} \, (= p_1 = p_2)$
Volume flow	$Q = 1400 \text{ l/min}$
Resistance coefficient - Inlet suction pipe	$\zeta_{IS} = 0{,}3$

Resistance coefficient - Inlet pressure pipe in the container	$\varsigma_{IC} = 0{,}7$
Resistance coefficient - 90°-pipe elbow	$\varsigma_{E} = 0{,}5$
Kinematic viscosity - water	$\nu = 10^{-6} \ \text{m}^2/\text{s}$
Absolut wall roughness of the pipes	$k = 0{,}5 \ \text{mm}$

Wanted:

The inlet pressure p_I and the outlet pressure p_O of the pump are to be calculated.

Solution:

1. To calculate the pressure at the pump inlet p_I the following equation is to be used

$$\frac{\rho}{2} v_1^2 + p_1 + \rho \cdot g \cdot z_1 = \frac{\rho}{2} v_I^2 + p_I + \rho \cdot g \cdot z_I + \Delta p_{V,1-I}$$

With $\Delta p_{V,1-I}$ the suction-side pressure losses are taken into account.

2. If this equation is changed to p_I we obtain

$$p_I = p_1 + \frac{\rho}{2}\left(v_1^2 - v_I^2\right) + \rho \cdot g \cdot (z_1 - z_I) - \Delta p_{V,1-I}$$

3. For the lower downstream water level 1 applies for the velocity $v_1 = 0$ and for the pressure $p_1 = p_{amb} = 1 \ \text{bar}$. The zero level is identical with the lower water level 1 (Figure 5.3). Therefore applies $z_1 = 0$. The geodetic height at the pump inlet is given by $z_I = -1{,}5 \ \text{m}$.

4. With $v_1 = 0$ and $z_1 = 0$ we obtain the equation for the pump inlet pressure

$$p_I = p_1 - \frac{\rho}{2} v_I^2 - \rho \cdot g \cdot z_I - \Delta p_{V,1-I}$$

Examples

5. For the (average) flow velocity at the pump inlet we obtain

$$v_1 = \frac{Q}{A} = \frac{Q}{d^2 \frac{\pi}{4}} = \frac{1400 \frac{1}{\min}}{0,1^2 \text{ m}^2 \frac{\pi}{4}} = \frac{1400 \frac{10^{-3} \text{ m}^3}{60 \text{ s}}}{0,1^2 \text{ m}^2 \frac{\pi}{4}} = 2,97 \text{ m/s}$$

6. The suction-side pressure losses $\Delta p_{V,1-I}$ consist of the portion which takes into account the losses in the suction tube and the portion that captures the losses at the inlet of the suction pipe.

$$\Delta p_{V,1-I} = \lambda_R \frac{l_S}{d} \frac{\rho}{2} v_1^2 + \zeta_{IS} \frac{\rho}{2} v_1^2$$

7. To determine the pipe friction coefficient λ_R, the *Reynolds*-number Re is required. It is

$$Re = \frac{v_1 \cdot d}{\nu} = \frac{2,97 \frac{\text{m}}{\text{s}} \cdot 0,1 \text{ m}}{10^{-6} \frac{\text{m}^2}{\text{s}}} = 297\,000$$

8. The friction coefficient λ_R is calculated using equation (2.62). For this purpose the values of $Re = 297\,000$ and $k/d = 0,5/100 = 0,005$ are required. This gives

$$\lambda_R = 0,03605$$

9. The losses in the $0,5 \text{ m}$ long suction pipe are

$$\lambda_R \frac{l_S}{d} \frac{\rho}{2} v_1^2 = 0,03065 \frac{0,5 \text{ m}}{0,1 \text{ m}} \frac{1000 \frac{\text{kg}}{\text{m}^3}}{2} 2,97^2 \frac{\text{m}^2}{\text{s}^2} = 675,90 \frac{\text{N}}{\text{m}^2}$$

10. The losses at the inlet of the suction pipe are

$$\zeta_{IS} \frac{\rho}{2} v_I^2 = 0,3 \frac{1000 \frac{kg}{m^3}}{2} 2,97^2 \frac{m^2}{s^2} = 1323,14 \frac{N}{m^2}$$

11. The suction-side pressure losses we thus obtain

$$\Delta p_{V,1-I} = 675,90 \frac{N}{m^2} + 1323,14 \frac{N}{m^2} = 1999,04 \frac{N}{m^2}$$

12. The pump inlet pressure can be calculated by using the equation described at point 4

$$p_I = 10^5 \frac{N}{m^2} - \frac{1000 \frac{kg}{m^3}}{2} 2,79^2 \frac{m^2}{s^2} - 1000 \frac{kg}{m^3} 9,81 \frac{m}{s^2} (-1,5 \text{ m}) - 1999,04 \frac{N}{m^2}$$

$$p_I = 108823,91 \frac{N}{m^2}$$

13. To calculate the pressure at the pump outlet p_O we use between the pump outlet O and the upstream water level 2 the equation

$$\frac{\rho}{2} v_O^2 + p_O + \rho \cdot g \cdot z_O = \frac{\rho}{2} v_2^2 + p_2 + \rho \cdot g \cdot z_2 + \Delta p_{V,O-2}$$

The pressure losses of the pressure side are taken into account with $\Delta p_{V,O-2}$.

14. If this equation is converted to p_O, we obtain

$$p_O = p_2 + \frac{\rho}{2} \left(v_2^2 - v_O^2 \right) + \rho \cdot g \left(z_2 - z_O \right) + \Delta p_{V,O-2}$$

15. For the upstream water level 2 applies $v_2 = 0$ for the velocity and for the pressure $p_2 = p_{amb} = 1 \text{ bar}$. Because the zero level acts on the down-

Examples

stream water level 1 $(z_1 = 0\,\text{m})$ applies $z_2 = 17\,\text{m}$ and $z_O = -1\,\text{m}$ (see Figure 5.3).

16. With $v_2 = 0$ we obtain the equation for the pump outlet pressure

$$p_O = p_2 - \frac{\rho}{2} v_O^2 + \rho \cdot g (z_2 - z_O) + \Delta p_{V,O-2}$$

17. For the cross sections O and I, the continuity equation is

$$A_O \cdot v_O = A_I \cdot v_I$$

18. The cross sections at the pump outlet O and the pump inlet I have the same size $(A_O = A_I)$. That means

$$v_O = v_I$$

19. For the calculation of the pressure losses $\Delta p_{V,O-2}$ on the pressure side we use the equation

$$\Delta p_{V,O-2} = \lambda_R \frac{l_D}{d} \frac{\rho}{2} v_O^2 + \zeta_E \frac{\rho}{2} v_O^2 + \zeta_{IC} \frac{\rho}{2} v_O^2$$

20. The losses in the 16 m and the 10 m long pipes are

$$\lambda_R \frac{l_D}{d} \frac{\rho}{2} v_O^2 = 0{,}03065 \, \frac{16\,\text{m} + 10\,\text{m}}{0{,}1\,\text{m}} \, \frac{1000 \frac{\text{kg}}{\text{m}^3}}{2} \, 2{,}97^2 \, \frac{\text{m}^2}{\text{s}^2} = 35146{,}9 \, \frac{\text{N}}{\text{m}^2}$$

21. The losses in the 90°-elbow are

$$\zeta_E \frac{\rho}{2} v_E^2 = 0{,}5 \, \frac{1000 \frac{\text{kg}}{\text{m}^3}}{2} \, 2{,}97^2 \, \frac{\text{m}^2}{\text{s}^2} = 2205{,}2 \, \frac{\text{N}}{\text{m}^2}$$

22. The losses at the inlet into the container are

$$\zeta_{IC} \frac{\rho}{2} v_0^2 = 0{,}7 \; \frac{1000 \frac{kg}{m^3}}{2} \; 2{,}97^2 \, \frac{m^2}{s^2} = 3087{,}3 \, \frac{N}{m^2}$$

23. The pressure losses $\Delta p_{V,O-2}$ on the pressure side are therefore

$$\Delta p_{V,O-2} = 35146{,}9 \, \frac{N}{m^2} + 2205{,}2 \, \frac{N}{m^2} + 3087{,}3 \, \frac{N}{m^2} = 40439{,}4 \, \frac{N}{m^2}$$

24. The pump outlet pressure is obtained by using the equation listed in point 16

$$p_O = 10^5 \, \frac{N}{m^2} - \frac{1000 \frac{kg}{m^3}}{2} \, 2{,}97^2 \, \frac{m^2}{s^2} +$$

$$1000 \, \frac{kg}{m^3} \, 9{,}81 \, \frac{m}{s^2} \left[17 \text{ m} - (-1 \text{ m}) \right] + 40439{,}4 \, \frac{N}{m^2}$$

$$p_O = 312608{,}95 \, \frac{N}{m^2}$$

25. The pressure difference at the pump is thus

$$\Delta p_P = p_O - p_I = 312608{,}95 \, \frac{N}{m^2} - 108823{,}91 \, \frac{N}{m^2} = 203785{,}04 \, \frac{N}{m^2}$$

$$\Delta p_P \approx 2{,}038 \, \text{bar}$$

Example 4:

From a tank, oil flows through a pipe in a higher lying container. The tank is under the pressure $p_{e,Tank}$ (Figure 5.4).

Examples

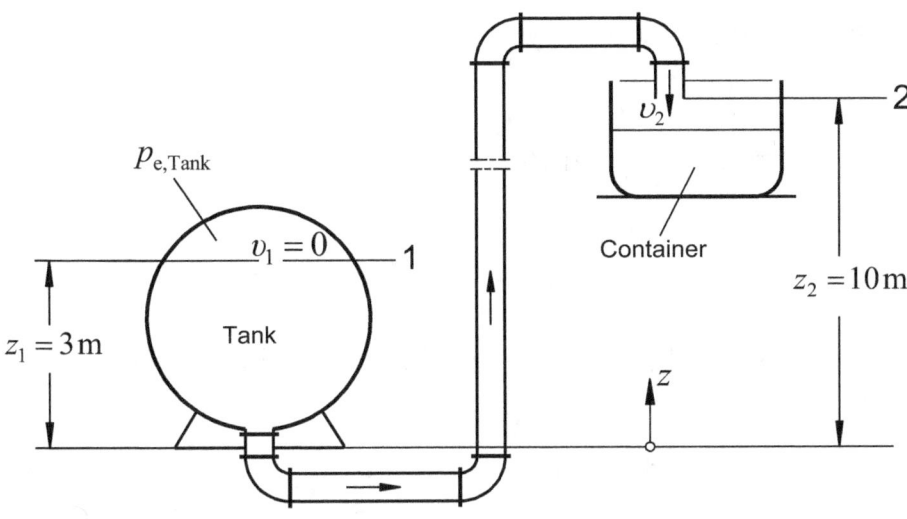

Figure 5.4: Oil flows from a tank into a container

For more data:

Internal pipe diameter	$d = 80$ mm
Volume flow	$Q = 100 \; \text{l}/\text{min}$
Density of the oil	$\rho = 950 \; \text{kg/m}^3$
Overall pressure losses	$\Delta p_{V,1-2} = 0{,}12$ bar

(Therein are also included caused by the 90°- elbows losses of pressure)

Wanted:

The pressure in the tank $p_{e,\text{Tank}}$ (overpressure) is to be calculated.

Solution:

1. To calculate the pressure in the tank $p_{e,\text{Tank}}$ the following equation has to be used

$$\frac{\rho}{2} v_1^2 + p_{\text{Tank}} + \rho \cdot g \cdot z_1 = \frac{\rho}{2} v_2^2 + p_{\text{amb}} + \rho \cdot g \cdot z_2 + \Delta p_{V,1-2}$$

The overall pressure losses caused by friction are taken into account by $\Delta p_{V,1-2}$

2. With $p_{Tank} = p_{e,Tank} + p_{amb}$ is obtained

$$\frac{\rho}{2} v_1^2 + p_{e,Tank} + p_{amb} + \rho \cdot g \cdot z_1 = \frac{\rho}{2} v_2^2 + p_{amb} + \rho \cdot g \cdot z_2 + \Delta p_{V,1-2}$$

3. If this equation is converted to $p_{e,Tank}$, we obtain with $v_1 = 0$

$$p_{e,Tank} = \frac{\rho}{2} v_2^2 + \rho \cdot g (z_2 - z_1) + \Delta p_{V,1-2}$$

4. The velocity v_2 is calculated by using the volume flow Q and the pipe inner diameter d to

$$v_2 = \frac{Q}{A_2} = \frac{Q}{d^2 \frac{\pi}{4}} = \frac{100 \frac{1}{\min}}{0,08^2 \text{ m}^2 \frac{\pi}{4}} = \frac{100 \frac{10^{-3} \text{ m}^3}{60 \text{ s}}}{0,08^2 \text{ m}^2 \frac{\pi}{4}} = 0,3316 \text{ m/s}$$

5. The overpressure is obtained by using the equation under point 3

$$p_{e,Tank} = \frac{950 \frac{\text{kg}}{\text{m}^3}}{2} 0,3316^2 \frac{\text{m}^2}{\text{s}^2} + 950 \frac{\text{kg}}{\text{m}^3} 9,81 \frac{\text{m}}{\text{s}^2} (10 \text{ m} - 3 \text{ m}) + 12000 \frac{\text{N}}{\text{m}^2}$$

$$p_{e,Tank} = 77288,73 \frac{\text{N}}{\text{m}^2} \approx 0,773 \text{ bar}$$

Example 5:

From a reservoir water flow through a pipe into a deeper-lying channel (Figure 5.5).

Examples

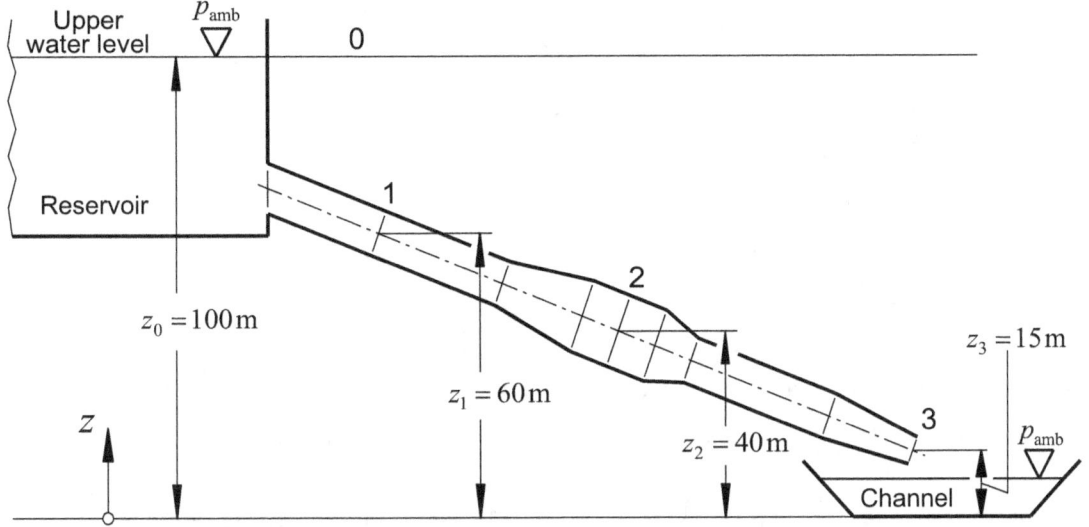

Figure 5.5: Water flows from a reservoir into a channel

For more data:

Internal pipe diameter at location 1	$d_1 = 200$ mm
Internal pipe diameter at location 2	$d_2 = 300$ mm
Internal pipe diameter at location 3	$d_3 = 50$ mm
Density of the water	$\rho = 1000$ kg/m³

Friction losses are remaining unconsidered (lossless flow).

Wanted:

The flow velocities and pressures (overpressures) of the cross-sections 1, 2 and 3 shall be calculated and the numerical values found to the equation (2.49) shall be illustrated.

Solution:

1. Based on equation (2.43) results for the cross-sections 1 and 3

$$z_0 + \frac{p_0}{\rho \cdot g} + \frac{v_0^2}{2 \cdot g} = z_3 + \frac{p_3}{\rho \cdot g} + \frac{v_3^2}{2 \cdot g}$$

2. With $p_0 = p_{e,0} + p_{amb}$ and $p_3 = p_{e,3} + p_{amb}$ we obtain

$$z_0 + \frac{p_{e,0}}{\rho \cdot g} + \frac{p_{amb}}{\rho \cdot g} + \frac{v_0^2}{2 \cdot g} = z_3 + \frac{p_{e,3}}{\rho \cdot g} + \frac{p_{amb}}{\rho \cdot g} + \frac{v_0^3}{2 \cdot g}$$

$$z_0 + \frac{p_{e,0}}{\rho \cdot g} + \frac{v_0^2}{2 \cdot g} = z_3 + \frac{p_{e,3}}{\rho \cdot g} + \frac{v_3^2}{2 \cdot g}$$

3. For the upper water level 0 applies for the velocity $v_0 = 0$ and for the pressure $p_{e,0} = 0$. Also applies $p_{e,3} = 0$ at the location 3.

4. Thus the flow velocity can be calculated at 3 (see equation at point 2), which is converted to v_3

$$v_3 = \sqrt{2 \cdot g (z_0 - z_3)} = \sqrt{2 \cdot 9{,}81 \frac{m}{s^2} (100\,m - 15\,m)} = 40{,}84\,m/s$$

5. The continuity equation for the cross-section 1 and 3 gives

$$A_1 \cdot v_1 = A_3 \cdot v_3$$

6. The (average) flow velocity at 1 is thus

$$v_1 = \frac{A_3}{A_1} v_3 = \left(\frac{d_3}{d_1}\right)^2 v_3 = \left(\frac{50\,mm}{200\,mm}\right)^2 40{,}84\,\frac{m}{s} = 2{,}55\,m/s$$

7. The continuity equation for the cross-section 2 and 3 gives

$$A_2 \cdot v_2 = A_3 \cdot v_3$$

8. The (average) flow velocity at 2 is thus

$$v_2 = \frac{A_3}{A_2} v_3 = \left(\frac{d_3}{d_2}\right)^2 v_3 = \left(\frac{50\,mm}{300\,mm}\right)^2 40{,}84\,\frac{m}{s} = 1{,}13\,m/s$$

Examples

9. The pressure (overpressure) at the location 1 is obtained from the equation

$$z_1 + \frac{p_{e,1}}{\rho \cdot g} + \frac{v_1^2}{2 \cdot g} = z_0 \text{ , which is converted to } p_{e,1}$$

$$p_{e,1} = \rho \cdot g(z_0 - z_1) - \rho \frac{v_1^2}{2} = \rho \left[g(z_0 - z_1) - \frac{v_1^2}{2} \right]$$

$$p_{e,1} = 1000 \frac{\text{kg}}{\text{m}^3} \left[9{,}81 \frac{\text{m}}{\text{s}^2} (100 \text{ m} - 60 \text{ m}) - \frac{2{,}55^2 \frac{\text{m}^2}{\text{s}^2}}{2} \right]$$

$$p_{e,1} = 389148{,}75 \frac{\text{N}}{\text{m}^2} \approx 3{,}9 \text{ bar}$$

10. The pressure (gauge pressure) at the location 2 is obtained from the equation

$$z_2 + \frac{p_{e,2}}{\rho \cdot g} + \frac{v_2^2}{2 \cdot g} = z_0 \text{ , which is converted to } p_{e,2}$$

$$p_{e,2} = \rho \cdot g(z_0 - z_2) - \rho \frac{v_2^2}{2} = \rho \left[g(z_0 - z_2) - \frac{v_2^2}{2} \right]$$

$$p_{e,2} = 1000 \frac{\text{kg}}{\text{m}^3} \left[9{,}81 \frac{\text{m}}{\text{s}^2} (100 \text{ m} - 40 \text{ m}) - \frac{1{,}13^2 \frac{\text{m}^2}{\text{s}^2}}{2} \right]$$

$$p_{e,2} = 587961{,}55 \frac{\text{N}}{\text{m}^2} \approx 5{,}9 \text{ bar}$$

11. If the calculated numerical values are used in equation (2.49) we obtain

$$z_0 = z_1 + \frac{p_{e,1}}{\rho \cdot g} + \frac{v_1^2}{2 \cdot g} = z_2 + \frac{p_{e,2}}{\rho \cdot g} + \frac{v_2^2}{2 \cdot g} = z_3 + \frac{p_{e,3}}{\rho \cdot g} + \frac{v_3^2}{2 \cdot g}$$

Location 1: $60 \text{ m} + \dfrac{389148,75}{1000 \cdot 9,81} \text{ m} + \dfrac{2,55^2}{2 \cdot 9,81} \text{ m} =$

$$(60 \text{ m} + 39,67 \text{ m} + 0,33 \text{ m} = 100 \text{ m})$$

Location 2: $40 \text{ m} + \dfrac{587961,55}{1000 \cdot 9,81} \text{ m} + \dfrac{1,13^2}{2 \cdot 9,81} \text{ m} =$

$$(40 \text{ m} + 59,93 \text{ m} + 0,07 \text{ m} = 100 \text{ m})$$

Location 3: $15 \text{ m} + \qquad 0 \text{ m} + \dfrac{40,84^2}{2 \cdot 9,81} \text{ m} =$

$$(15 \text{ m} + \qquad 0 \text{ m} + 85 \text{ m} = 100 \text{ m})$$

12. In Figure 5.6 these numerical values are illustrated as lines

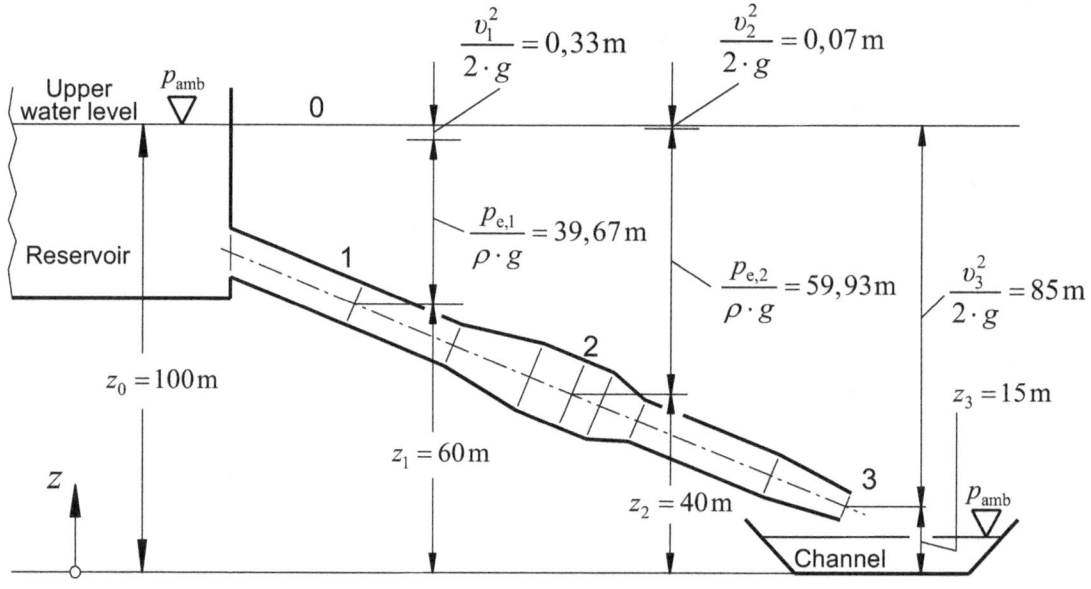

Figure 5.6: Illustration of numerical values

Examples 79

> NOTE: In Figure 5.6, the values $v_1^2/2 \cdot g$ and $v_2^2/2 \cdot g$ shown as lines are drawn enlarged; a factor of 10 has been used.

Example 6:

The Figure 5.7 shows a hydraulic press with pressure transmission in additional. The press is used as a metal forming machine tool. The maximum force that can be generated by the press is $F_3 = 500 \text{ kN}$ with a stroke of $s_3 = 5 \text{ mm}$.

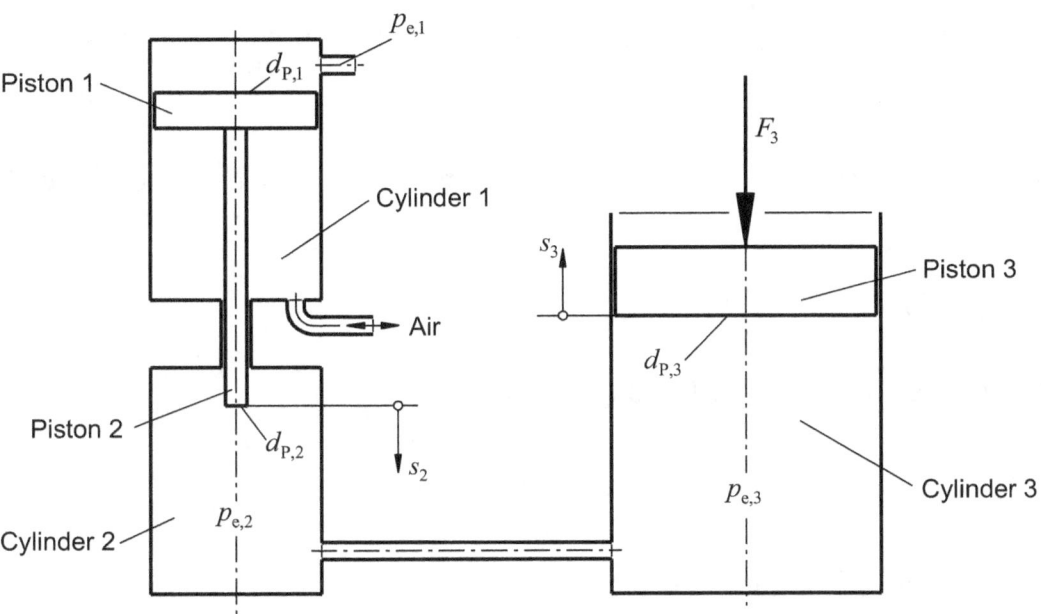

Figure 5.7: Hydraulic press with additional pressure transmission

For more data:

Diameter - piston 1	$d_{P,1} = 200 \text{ mm}$
Diameter - piston 2	$d_{P,2} = 50 \text{ mm}$
Fluid pressure (overpressure)	$p_{e,1} = 4 \text{ bar}$

It is assumed that the pistons are guided without friction and have no leakage.

Examples

Wanted:

It shall be calculated: The diameter of the piston 3 $d_{P,3}$ and the stroke of the piston 2 s_2.

Solution:

1. When applying the relevant equation (2.26) is obtained for the pressure (overpressure) in the cylinder 2 is obtained as follows

$$p_{e,2} = p_{e,1} \frac{A_{P,1}}{A_{P,2}} = p_{e,1} \frac{d_{P,1}^2 \frac{\pi}{4}}{d_{P,2}^2 \frac{\pi}{4}} = p_{e,1} \left(\frac{d_{P,1}}{d_{P,2}} \right)^2 = 4 \text{ bar} \left(\frac{200 \text{ mm}}{50 \text{ mm}} \right)^2 = 64 \text{ bar}$$

2. The pressure $p_{e,2}$ prevails according to the law of *Pascal* in cylinder 3 too. It is

$$p_{e,3} = p_{e,2} = 64 \text{ bar}$$

3. The equilibrium at the piston 3 supplies

$$p_{e,3} \cdot A_{P,3} = p_{e,3} \cdot d_{P,3}^2 \frac{\pi}{4} = F_3 = 500 \text{ kN} = 500\,000 \text{ N}$$

4. This leads to the diameter of the piston 3

$$d_{P,3} = \sqrt{\frac{4}{\pi} \frac{F_3}{p_{e,3}}} = 1{,}1284 \sqrt{\frac{F_3}{p_{e,3}}} = 1{,}1284 \sqrt{\frac{500\,000 \text{ N}}{64 \cdot 10^5 \frac{\text{N}}{\text{m}^2}}} = 0{,}315 \text{ m} = 315 \text{ mm}$$

5. To lift the piston 3 a volume V_3 must be pressed into cylinder 3

$$V_3 = \frac{\pi}{4} d_{P,3}^2 \cdot s_3 = \frac{\pi}{4} 315^2 \text{ mm}^2 \cdot 5 \text{ mm} = 389\,656 \text{ mm}^3$$

Examples

6. The liquid volume V_3 has to be displaced from the cylinder 2. So applies

$$V_2 = s_2 \cdot A_{P,2} = s_2 \cdot d_{P,2}^2 \frac{\pi}{4} = V_3 = 389\,656 \text{ mm}^3$$

7. Thus the stroke of the piston 2 results in

$$s_2 = \frac{4}{\pi} \frac{V_2}{d_{P,2}^2} = \frac{4}{\pi} \frac{389\,656 \text{ mm}^3}{50^2 \text{ mm}^2} = 198,5 \text{ mm}$$

Example 7:

Figure 5.8 shows two cylinders which are connected by a pipe. Both pistons have no friction. The hydraulic fluid is sealed without leakage to the atmosphere through the piston seals. The force $F_1 = 100$ N acts on piston 1.

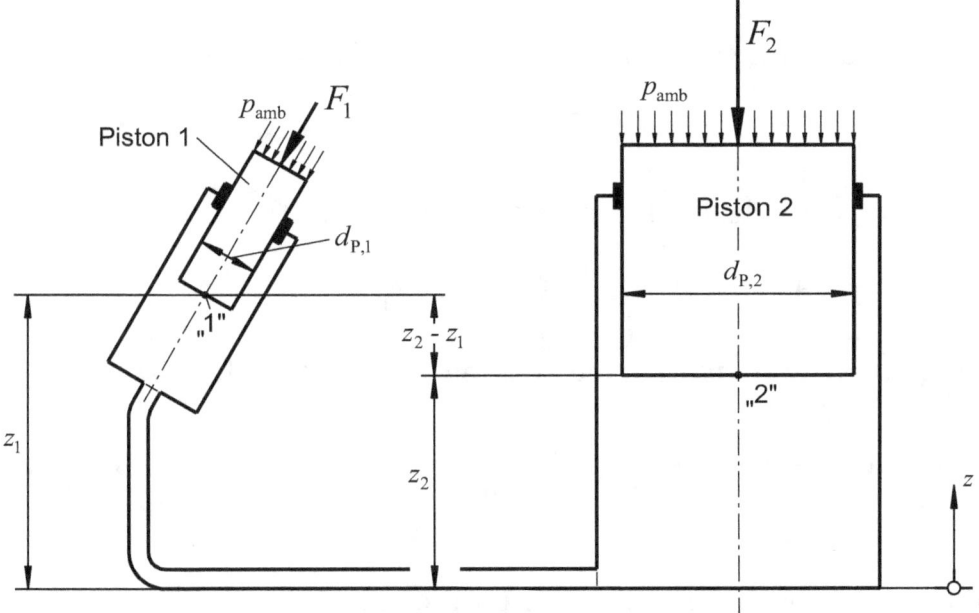

Figure 5.8: Two cylinders which are connected by a pipe

For more data:

Diameter - Piston 1 $\hspace{5cm} d_{P,1} = 125 \text{ mm}$

Diameter - Piston 2 $\qquad d_{P,2} = 500$ mm

Density of the fluid $\qquad \rho = 900$ kg/m^3

Geodetic level - Position „1" $\qquad z_1 = 600$ mm

Geodetic level - Position „2" $\qquad z_2 = 520$ mm

Wanted:

The force F_2 is to be calculated, which acts on the piston 2 (the gravity pressure of the hydraulic oil shall be considered).

Solution:

1. The equilibrium of forces on the piston 1 delivers the equation

$$F_1 + p_{amb} \cdot A_1 = p_{abs,1} \cdot A_1$$

2. The pressure (absolute pressure) at position „1" is expressed by

$$p_{abs,1} = p_{e,1} + p_{amb}$$

3. Used in the equation listed under point 1, we obtain

$$F_1 + p_{amb} \cdot A_1 = (p_{e,1} + p_{amb}) A_1 = p_{e,1} \cdot A_1 + p_{amb} \cdot A_1$$

4. The pressure (overpressure) at the position „1" will therefore be

$$p_{e,1} = \frac{F_1}{A_1} = \frac{F_1}{d_{P,1}^2 \frac{\pi}{4}} = \frac{100 \text{ N}}{0{,}125^2 \text{ m}^2 \frac{\pi}{4}} = 8149 \frac{\text{N}}{\text{m}^2} = 8149 \cdot 10^{-5} \text{ bar} = 0{,}08149 \text{ bar}$$

5. The pressure (absolut pressure) at the position „2" is

$$p_{abs,2} = p_{e,2} + p_{amb}$$

6. The pressure (absolut pressure) at the position „2" is also

$$p_{abs,2} = p_{abs,1} + \rho \cdot g\,(z_1 - z_2)$$

7. The equations under points 5 and 6 are set equal. Then we get

$$p_{e,2} + p_{amb} = p_{abs,1} + \rho \cdot g\,(z_1 - z_2)$$

8. With $p_{abs,1} = p_{e,1} + p_{amb}$ (see point 2) we get

$$p_{e,2} + p_{amb} = p_{e,1} + p_{amb} + \rho \cdot g\,(z_1 - z_2)$$

9. For the pressure at position „2" the following equation applies

$$p_{e,2} = p_{e,1} + \rho \cdot g\,(z_1 - z_2)$$

10. The force equilibrium at the piston 2 provides the equation

$$F_2 + p_{amb} \cdot A_2 = p_{abs,2} \cdot A_2 = (p_{e,2} + p_{amb})\,A_2 = p_{e,2} \cdot A_2 + p_{amb} \cdot A_2$$

11. If the equation is used found for $p_{e,2}$ (see point 9), we get

$$F_2 + p_{amb} \cdot A_2 = \left[p_{e,1} + \rho \cdot g\,(z_1 - z_2) \right] A_2 + p_{amb} \cdot A_2$$

12. Now we can calculate the required force

$$F_2 = \left[p_{e,1} + \rho \cdot g\,(z_1 - z_2) \right] A_2 = \left[p_{e,1} + \rho \cdot g\,(z_1 - z_2) \right] d_{P,2}^2 \frac{\pi}{4}$$

$$F_2 = \left[8149\,\frac{N}{m^2} + 900\,\frac{kg}{m^3}\,9{,}81\,\frac{m}{s^2}\,(0{,}6\,m - 0{,}52\,m) \right] 0{,}5^2\,m^2\,\frac{\pi}{4} = 1739\,N$$

Example 8:

Figure 5.9 shows a hydraulic cylinder to whose piston rod a rope is fastened. The rope is redirected via a rope pulley. At the end of the rope a weight G is attached.

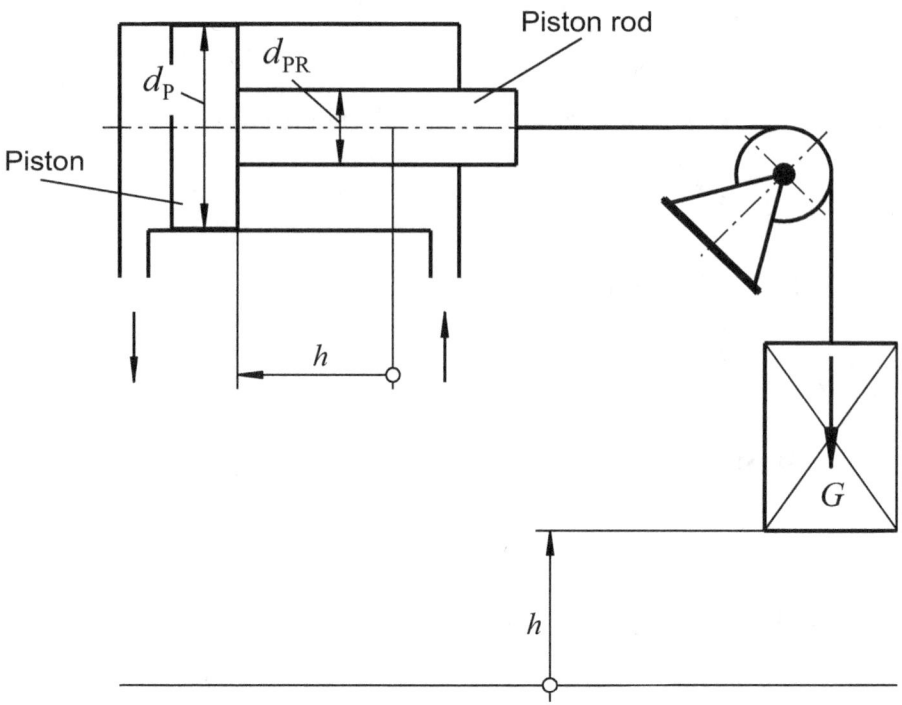

Figure 5.9: Hydraulic cylinder to whose piston rod a rope is fastened

Data of the hydraulic cylinder:

Hydraulic- mechanical efficiency - piston side	$\eta_{hm,P} = 0{,}98$
Hydraulic-mechanical efficiency - piston rod side	$\eta_{hm,PR} = 0{,}96$
Volumetric efficiency	$\eta_v = 1$
Diameter - Piston	$d_P = 80$ mm
Diameter-Piston rod	$d_{PR} = 45$ mm
Piston side pressure during the stroke	$p_P = 2$ bar

Wanted:

It should be calculated: The piston rod-side pressure in the hydraulic cylinder and the effective volume flow of the hydraulic pump. The weight $G = 40\,000$ N is to be lifted in the time $t = 12$ s to a height of $h = 800$ mm.

Examples

Acceleration forces and pulley friction are ignored.

Solution:

1. Figure 5.10 illustrates the forces on piston and piston rod, when the piston rod retracts.

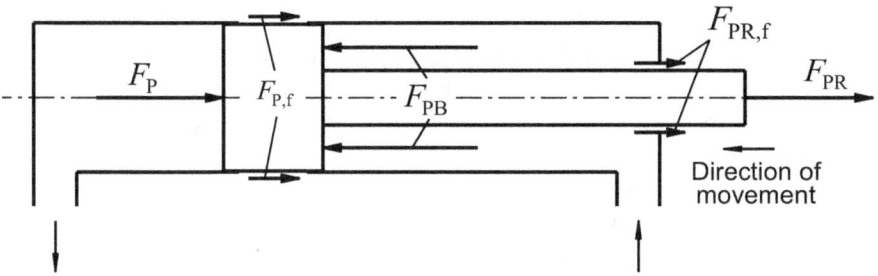

Figure 5.10: Situation of forces on the piston and piston rod of a so-called differential hydraulic cylinder - retraction

2. The equilibrium of forces on the piston and piston rod provides the following equation

$$F_P + F_{P,f} - F_{PB} + F_{PR,f} + F_{PR} = 0$$

3. The conversion of this equation with respect to the piston rod force F_{PR} leads to

$$F_{PR} = -F_P - F_{P,f} + F_{PB} - F_{PR,f}$$

$$F_{PR} = -(F_P + F_{P,f}) + (F_{PB} - F_{PR,f})$$

$$F_{PR} = -F_P\left(1 + \frac{F_{P,f}}{F_P}\right) + F_{PB}\left(1 - \frac{F_{PR,f}}{F_{PB}}\right)$$

$$F_{PR} = F_{PB}\left(1 - \frac{F_{PR,f}}{F_{PB}}\right) - F_P\left(1 + \frac{F_{P,f}}{F_P}\right)$$

4. The hydraulic-mechanical efficiencies for the piston rod side and the piston side have to be defined here as follows

Piston rod side: $\eta_{hm,PR} = 1 - \dfrac{F_{PR,f}}{F_{PB}}$ Piston side: $\eta_{hm,P} = \dfrac{1}{1 + \dfrac{F_{P,f}}{F_P}}$

5. This gives the equation, which is important for further procession of this example

$$F_{PR} = F_{PB} \cdot \eta_{hm,PR} - \dfrac{F_P}{\eta_{hm,P}}$$

6. Taking into account pressures and areas we will receive

$$F_{PR} = p_{PB} \cdot A_{PB} \cdot \eta_{hm,PR} - \dfrac{p_P \cdot A_P}{\eta_{hm,P}}$$

7. This equation is converted to p_{PB}. This gives

$$p_{PB} = \dfrac{F_{PR} + \dfrac{p_P \cdot A_P}{\eta_{hm,P}}}{A_{PB} \cdot \eta_{hm,PR}}$$

8. The piston area A_P is given by

$$A_P = d_P^2 \dfrac{\pi}{4} = 80^2 \, mm^2 \, \dfrac{\pi}{4} = 5\,026{,}54 \, mm^2$$

9. For the piston area on the piston rod side A_{PB} is obtained

$$A_{PB} = A_P - d_{PR}^2 \dfrac{\pi}{4} = 5\,026{,}54 \, mm^2 - 45^2 \, mm^2 \, \dfrac{\pi}{4} = 3\,436{,}11 \, mm^2$$

Examples

10. The given data are $F_{PR} = 40\,000$ N, $\eta_{hm,P} = 0,98$, $\eta_{hm,PR} = 0,96$, $p_{PB} = 2$ bar. With these data we obtain for the piston rod side pressure

$$p_{PB} = \frac{40\,000\text{ N} + \dfrac{2\text{ bar}\cdot 5026,54\text{ mm}^2}{0,98}}{3436,11\text{ mm}^2 \cdot 0,96} = \frac{40\,000\text{ N} + \dfrac{2\cdot 0,1\text{ N/mm}^2 \cdot 5026,54\text{ mm}^2}{0,98}}{3436,11\text{ mm}^2 \cdot 0,96}$$

$p_{PB} = 12,44\text{ N/mm}^2 = 12,44 \cdot 10$ bar

$p_{PB} = 124,4$ bar

11. The piston rod side inflowing volume with $h = 800$ mm is

$V_{PB} = h \cdot A_{PB} = 800\text{ mm} \cdot 3436,11\text{ mm}^2$

$V_{PB} = 2\,748\,888\text{ mm}^3 = 2\,748\,888 \cdot 10^{-6}\text{ dm}^3 \approx 2,75\text{ l}$

12. The weight is lifted in the time $t = 12$ s. The effective volume flow of the hydraulic pump is

$$Q_{e,P} = \frac{V_{PB}}{t} = \frac{2,75\text{ l}}{12\text{ s}} = \frac{2,75\text{ l}}{12\,\dfrac{1}{60}\text{ min}} = \frac{2,75\text{ l}\cdot 60}{12\text{ min}} = 13,75\text{ l/min}$$

Example 9:

An oil-filled straight pipe is considered in different pressure states (Figure 5.11). State "above": the oil is under the atmospheric pressure $p_{amb} = 1$ bar. State "middle" and state "below": the oil is under the overpressure $p_e = 200$ bar.

State "above"

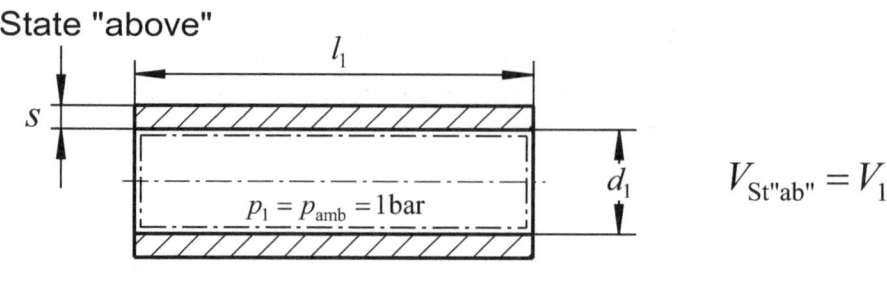

State "middle" (incompressible)

State "below" (compressible)

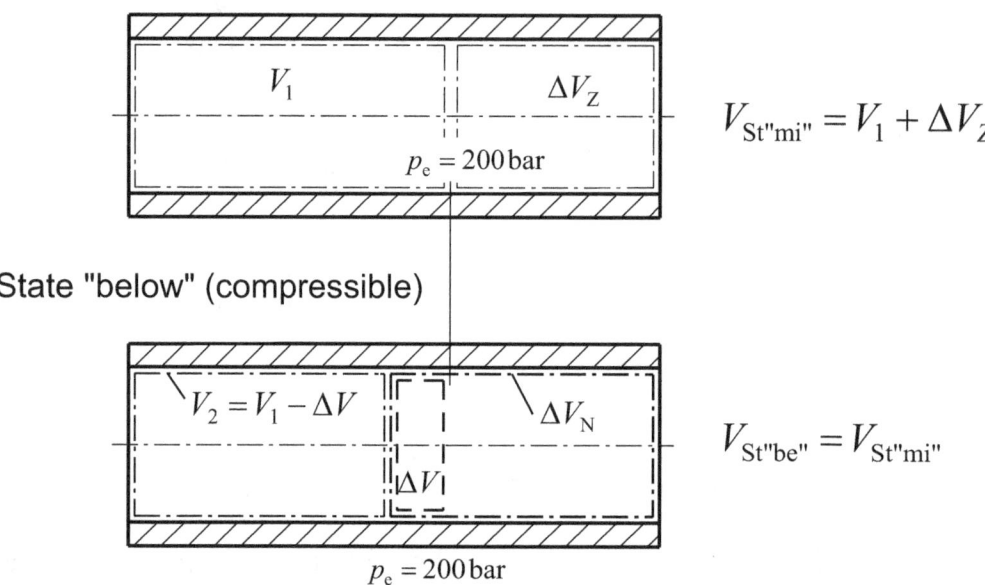

Figure 5.11: An oil-filled pipe in different states

For more data:

Average compression module	$K_S = 20\,000$ bar
Poisson constant - steel	$m = 3,3$
Modulus of	$E = 210\,000$ N/mm^2
Inner diameter of the pipe - state "above"	$d_1 = 20$ mm

Examples

Length of the pipe - state "above" $l_1 = 5\,\text{m}$

Wall thickness of the pipe $s = 3\,\text{mm}$

Wanted:

Taking into account the compressibility of the oil and the elastic behavior of the pipe following question is to be answered: How big is the oil volume which needs to be added to the complete filling of the pipe? For a better understanding is to look at Figure 5.11.

Solution:

1. **State "above"**: the oil inside the pipe is under the atmospheric pressure $p_{amb} = 1\,\text{bar}$. In this state the volume is $V_{St"ab"} = V_1$.

2. **State "middle"**: the oil **assumed as incompressible** is in this state under the pressure $p_e = 200\,\text{bar}$ (overpressure). The pipe has widened and lengthened due to its elastic behavior. The volume in this state is $V_{St"mi"} = V_1 + \Delta V_Z$.

3. **State "below"**: the oil volume V_1 is reduced as a result of the compressibility to $V_2 = V_1 - \Delta V$. The searched volume is then obtained using the equation $\Delta V_N = \Delta V_Z + \Delta V$.

4. The oil volume in state "above" is

$$V_{St"ab"} = V_1 = d_1^2 \frac{\pi}{4} l_1 = 20^2\,\text{mm}^2 \frac{\pi}{4} 5000\,\text{mm} = 1{,}5708 \cdot 10^6\,\text{mm}^3 = 1570{,}8\,\text{cm}^3$$

5. The laws of the theory of elasticity allows calculating the increase in volume of a pipe with circular cross section, when the pressure in the pipe is higher than the atmospheric pressure with the following equation

$$\Delta V_Z = V_1 \left(\frac{5}{4} - \frac{1}{m} \right) \frac{d_1 \cdot \Delta p}{s \cdot E}$$

Herein Δp is the difference of the absolut pressures in the two pressure states of the oil. Therefore is $\Delta p = 200$ bar.

$$\Delta V_Z = 1570{,}8 \text{ cm}^3 \left(\frac{5}{4} - \frac{1}{3{,}3}\right) \frac{20 \text{ mm} \cdot 200 \text{ bar}}{3 \text{ mm} \cdot 210000 \frac{\text{N}}{\text{mm}^2}}$$

$$\Delta V_Z = 1570{,}8 \text{ cm}^3 \left(\frac{5}{4} - \frac{1}{3{,}3}\right) \frac{20 \text{ mm} \cdot 200 \cdot 10^5 \frac{\text{N}}{\text{m}^2}}{3 \text{ mm} \cdot 210000 \frac{\text{N}}{\text{mm}^2}}$$

$$\Delta V_Z = 1570{,}8 \text{ cm}^3 \left(\frac{5}{4} - \frac{1}{3{,}3}\right) \frac{20 \text{ mm} \cdot 200 \cdot 10^5 \frac{\text{N}}{10^6 \text{ mm}^2}}{3 \text{ mm} \cdot 210000 \frac{\text{N}}{\text{mm}^2}}$$

$$\Delta V_Z = 1570{,}8 \text{ cm}^3 \left(\frac{5}{4} - \frac{1}{3{,}3}\right) \frac{20 \cdot 200 \cdot 10^{-1}}{3 \cdot 210000}$$

$$\Delta V_Z = 0{,}94444 \text{ cm}^3$$

6. To calculate the volume decrease ΔV due to the influence of the compressibility reference is made to equation (2.113). This is

$$V_2 = V_1 \left(1 - \frac{\Delta p}{K_S}\right)$$

For V_2 is inserted $V_2 = V_1 - \Delta V$. By rearranging to ΔV we obtain the equation for the volume decrease

$$\Delta V = V_1 \frac{\Delta p}{K_S}$$

$$\Delta V = 1570{,}8 \text{ cm}^3 \; \frac{200 \, \text{bar}}{20\,000 \, \text{bar}} = 15{,}708 \, \text{cm}^3$$

7. The recharge volume $\Delta V_N = \Delta V_Z + \Delta V$ is

$$\Delta V_N = 0{,}94444 \text{ cm}^3 + 15{,}708 \, \text{cm}^3 = 16{,}652 \, \text{cm}^3$$

That`s about $1{,}06\,\%$ of the state "above" volume $V_1 = 1570{,}8 \, \text{cm}^3$.

Example 10:

For a hydraulically operated lift system the main design data of hydraulic cylinder and hydraulic pump are to be determined. The load on the piston rod of the hydraulic cylinder during the stroke is $F_{PR} = 58\,000 \text{ N}$.

Data - Hydraulic cylinder (differential cylinder):

Maximum pressure - piston side	$p_{P,\max} = 80 \text{ bar}$
Area ratio $\varphi = A_P / A_{PB} \approx 1{,}6$	
Lifting speed	$v_{\text{Lift}} = 100 \text{ mm/s}$
Hydraulic- mechanical efficiency - piston side	$\eta_{\text{hm,PS}} = 0{,}97$
Hydraulic-mechanical efficiency - piston rod side	$\eta_{\text{hm,PB}} = 0{,}95$
Volumetric efficiency	$\eta_{\text{v,HC}} = 1$

Data - Hydraulic pump:

Hydraulic- mechanical efficiency	$\eta_{\text{hm,P}} = 0{,}97$
Volumetric efficiency	$\eta_{\text{v,P}} = 0{,}94$

Other data:

Pressure loss between pump and hydraulic cylinder $\Delta p_{L,P-HC} = 4 \text{ bar}$

Pressure loss in the return pipe $\Delta p_{L,RP} = 1 \text{ bar}$

Atmospheric pressure $p_{amb} = 1 \text{ bar}$

Wanted:

It is to be calculated: the inner diameter of the hydraulic cylinder, the effective area ratio of the hydraulic cylinder, the piston-side pressure in the hydraulic cylinder, the effective volume flow rate of the hydraulic pump, the mechanical drive power of the hydraulic pump and the overall efficiency of the system during the stroke (acceleration forces are ignored).

Solution:

1. Figure 5.12 illustrates the forces on piston and piston rod, when the piston rod extends.

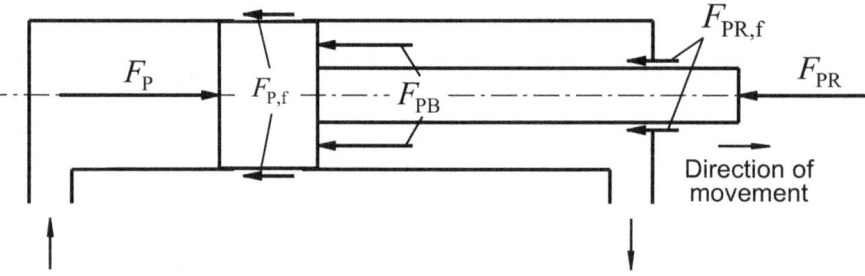

Figure 5.12: Situation of forces on the piston and piston rod of a so-called differential hydraulic cylinder - piston rod extends

2. The equilibrium of forces on the piston and piston rod supplies the following equation

$$F_P - F_{P,f} - F_{PB} - F_{PR,f} - F_{PR} = 0$$

Examples

3. The conversion of this equation with respect to the piston rod force F_{PR} leads to

$$F_{PR} = F_P - F_{P,f} - F_{PB} - F_{PR,f}$$

$$F_{PR} = \left(F_P - F_{P,f}\right) - \left(F_{PB} + F_{PR,f}\right)$$

$$F_{PR} = F_P\left(1 - \frac{F_{P,f}}{F_P}\right) - F_{PB}\left(1 + \frac{F_{PR,f}}{F_{PB}}\right)$$

4. The hydraulic-mechanical efficiencies for the piston side and the piston rod side have to be defined here as follows

Piston side: $\eta_{hm,PS} = 1 - \dfrac{F_{P,f}}{F_P}$ Piston rod side: $\eta_{hm,PB} = \dfrac{1}{1 + \dfrac{F_{PR,f}}{F_{PB}}}$

5. This provides the equation, which is important for the further procession of this example

$$F_{PR} = F_P \cdot \eta_{hm,PS} - \frac{F_{PB}}{\eta_{hm,PB}}$$

6. This equation is used for the calculation of the inner diameter of the hydraulic cylinder

$$F_{PR} = F_P \cdot \eta_{hm,PS} - \frac{F_{PB}}{\eta_{hm,PB}} = p_P \cdot A_P \cdot \eta_{hm,PS} - \frac{p_{PB} \cdot A_{PB}}{\eta_{hm,PB}}$$

7. Using $A_{PB} = A_P/\varphi$ and $p_P = p_{P,max}$ we obtain by converting the equation for calculation of the piston area

$$A_\mathrm{P} = \frac{F_\mathrm{PR}}{p_\mathrm{P,max} \cdot \eta_\mathrm{hm,PS} - \dfrac{p_\mathrm{PB}}{\varphi \cdot \eta_\mathrm{hm,PB}}}$$

8. The piston rod-side pressure is given by

$$p_\mathrm{PB} = p_\mathrm{amb} + \Delta p_\mathrm{L,RP} = 1\,\mathrm{bar} + 1\,\mathrm{bar} = 2\,\mathrm{bar}$$

9. For the under point 7 listed equation it can be written

$$A_\mathrm{P} = \frac{F_\mathrm{PR}}{p_\mathrm{P,max} \cdot \eta_\mathrm{hm,PS} - \dfrac{p_\mathrm{amb} + \Delta p_\mathrm{L,RP}}{\varphi \cdot \eta_\mathrm{hm,PB}}}$$

10. With the above data, we obtain for the piston area

$$A_\mathrm{P} = \frac{58\,000\,\mathrm{N}}{80\,\mathrm{bar} \cdot 0{,}97 - \dfrac{2\,\mathrm{bar}}{1{,}6 \cdot 0{,}95}} = \frac{58\,000\,\mathrm{N}}{80 \cdot 0{,}1\,\mathrm{N/mm^2} \cdot 0{,}97 - \dfrac{2 \cdot 0{,}1\,\mathrm{N/mm^2}}{1{,}6 \cdot 0{,}95}}$$

$$A_\mathrm{P} = 7603\,\mathrm{mm^2}$$

11. With $A_\mathrm{P} = D^2 \dfrac{\pi}{4}$ the inner diameter of the hydraulic cylinder is obtained

$$D = \sqrt{\frac{4 \cdot A_\mathrm{P}}{\pi}} = \sqrt{\frac{4 \cdot 7603\,\mathrm{mm^2}}{\pi}} \approx 98\,\mathrm{mm}$$

12. It is chosen a **standardized** hydraulic cylinder with an inner diameter $D = 100\,\mathrm{mm}$ and piston rod diameter $d_\mathrm{PB} = 63\,\mathrm{mm}$. With these data the actual area ratio of the hydraulic cylinder is

Examples

$$\varphi = \frac{A_P}{A_{PB}} = \frac{A_P}{A_P - d_{PB}^2 \frac{\pi}{4}} = \frac{D^2 \frac{\pi}{4}}{D^2 \frac{\pi}{4} - d_{PB}^2 \frac{\pi}{4}} = \frac{D^2}{D^2 - d_{PB}^2}$$

$$\varphi = \frac{100^2 \text{ mm}^2}{100^2 \text{ mm}^2 - 63^2 \text{ mm}^2} = 1,658$$

13. The piston-side pressure in the hydraulic cylinder is obtained by using the equation listed in point 6

$$p_P = \frac{F_{PR} + \dfrac{p_{PB} \cdot A_{PB}}{\eta_{hm,PB}}}{A_P \cdot \eta_{hm,PS}} = \frac{F_{PR} + \dfrac{p_{PB} \cdot A_P}{\varphi \cdot \eta_{hm,PB}}}{A_P \cdot \eta_{hm,PS}} = \frac{F_{PR} + \dfrac{p_{PB} \frac{\pi}{4} D^2}{\varphi \cdot \eta_{hm,PB}}}{D^2 \frac{\pi}{4} \eta_{hm,PS}}$$

$$p_P = \frac{58\,000 \text{ N} + \dfrac{2 \cdot 0,1 \text{ N/mm}^2 \frac{\pi}{4} 100^2 \text{ mm}^2}{1,658 \cdot 0,95}}{100^2 \text{ mm}^2 \frac{\pi}{4} 0,97} = 7,74 \frac{\text{N}}{\text{mm}^2} = 77,4 \text{ bar}$$

14. Thus, the pressure at the pump outlet is (taking into account the pressure loss between hydraulic pump and hydraulic cylinder)

$$p_{O,P} = p_P + \Delta p_{L,P-HC} = 77,4 \text{ bar} + 4 \text{ bar} = 81,4 \text{ bar}$$

15. The effective flow rate of the hydraulic pump is

$$Q_{e,P} = A_P \cdot v_{Lift} = D^2 \frac{\pi}{4} v_{Lift}$$

$$Q_{e,P} = 0,1^2 \text{m}^2 \frac{\pi}{4} 0,1 \text{ m/s} = 0,0007854 \text{ m}^3/\text{s}$$

$Q_{e,P} = 0,0007854 \cdot 60000 \text{ l/min} = 47,12 \text{ l/min}$

16. With the overall efficiency $\eta_{t,P} = \eta_{hm,P} \cdot \eta_{v,P} = 0,97 \cdot 0,94 = 0,9118$ the mechanical drive power of the hydraulic pump during the stroke is

$$P_{m,P} = \frac{Q_{e,P} \cdot \Delta p_P}{\eta_{t,P}} = \frac{Q_{e,P} \, (p_{O,P} - p_{I,P})}{\eta_{t,P}} = \frac{Q_{e,P} \, (p_{O,P} - p_{amb})}{\eta_{t,P}}$$

$$P_{m,P} = \frac{0,0007854 \text{ m}^3/\text{s} \, (81,4 \text{ bar} - 1 \text{ bar})}{0,9118}$$

$$P_{m,P} = \frac{0,0007854 \text{ m}^3/\text{s} \cdot 80,4 \cdot 10^5 \, \frac{\text{N}}{\text{m}^2}}{0,9118} = 6925,44 \text{ W} \approx 7 \text{ kW}$$

17. The overall efficiency of the hydraulic system during the stroke is

$$\eta_{System} = \frac{F_{PB} \cdot v_{Lift}}{P_{m,P}} = \frac{58000 \text{ N} \cdot 0,1 \text{ m/s}}{6925,44 \, \frac{\text{Nm}}{\text{s}}} \approx 0,84$$

Sources of Literature

NOTE: As mentioned in the preface, the book presented here is based essentially on chapter 2 of the book that the author has written in the German language. Below some literature sources are listed, which were also used for writing the book "**Grundlagen der Hydraulik**". On a translation into the English language has been omitted.

Allweiler: Innovative Pumpentechnik, Prospektmappe, Radolfzell

Backè, W., Hahmann, W.: Kennlinien und Kennlinienfelder hydrostatischer Getriebe, VDI-Berichte Nr. 138, Düsseldorf: VDI-Verlag 1969

Sources of Literature

Backè, W.: Systematik der hydraulischen Widerstandsschaltungen in Ventilen und Regelkreisen, Mainz: Krausskopf-Verlag 1974

Bauer, G.: Ölhydraulik, Stuttgart: B. G. Teubner 1998

Bienert, H. W.: Planung ölhydraulischer Anlagen, Ölhydraulik und Pneumatik 6 (Heft Nr. 3) 1962

Bosch: Hydraulik in Theorie und Praxis, Autor: Werner Götz, Herausgeber: Robert Bosch GmbH Geschäftsbereich Automatisierungstechnik Schulung (AT/VSZ) 1997

Chaimowitsch, E. M.: Ölhydraulik, Berlin: VEB Verlag Technik 1957

Dietterle, H.: Druckflüssigkeiten, Sonderdruck aus Krauskopf-Taschenbücher Ölhydraulik und Pneumatik, Band 1: Grundlagen der Ölhydraulik, Mainz: Krausskopf-Verlag

Dürr, A., Wachter, O.: Hydraulik in Werkzeugmaschinen, München: Carl Hanser Verlag

Eberthäuser, H., Helduser, S.: Fluidtechnik von A bis Z, Mainz: Vereinigte Fachverlage 1995

Eck, B.: Technische Strömungslehre, Berlin, Heidelberg, NewYork: Springer-Verlag 1978

Findeisen, F. u. D.: Ölhydraulik, Berlin, Heidelberg, NewYork: Springer-Verlag 1978

Fister, W.: Fluidenergiemaschinen, Berlin, Heidelberg, NewYork: Springer-Verlag 1984

Foitzik, B.: Filtertechnologie für Hydrauliksysteme, Landsberg/Lech: Verlag Moderne Industrie 1996

Guillon, M.: Hydrostatische Regelkreise und Servosteuerungen, Grundlagen, Berechnungen und Anwendungen, München: Carl Hanser Verlag 1968

Hahn: Katalog Standard-Hydrozylinder, Sprockhövel

Halberg: Hydraulische Grundlagen für den Entwurf von Kreiselpumpenanlagen (Teil 1), Ludwigshafen/Rhein

Hauhinco: Radialkolbenpumpen (div. Druckschriften), Sprockhövel

Hänchen: Hydraulik-Zylinder, Prospektkennziffer ADWERB – HH 2518/543210, Ostfildern

Hänchen: Hydraulik-Zylinder, Ratio-Clamp Prospektkennziffer ADWERB – HH 2513/543210, Ostfildern

Herning, F.: Stoffströme in Rohrleitungen, Düsseldorf: VDI-Verlag 1966

HYDAC: Rückschlagventile hydraulisch entsperrbar ERVE, Prospekt Nr. 5.172.5/8.94 Katalog 01 Rubrik 09, Sulzbach/Saar

HYDAC: Drosselventile und Drosselrückschalagventile DVP, DRVP, Prospekt Nr. 5.120.0/4.96 Katalog 01 Rubrik 08, Sulzbach/Saar

HYDAC: Wechselventile WVT, Prospekt Nr. 5.178.2/5.96 Katalog 01 Rubrik 09, Sulzbach/Saar

HYDAC: Druckbegrenzungsventile DB10, Prospekt Nr. 5.167.1/8.95 Katalog 01 Rubrik 07, Sulzbach/Saar

HYDAC: Rohrbruchsicherungen RBE, Prospekt Nr. 5.174.5/12.94 Katalog 01 Rubrik 09, Sulzbach/Saar

HYDAC: 2-Wege-Stromregelventile SRE, SR5E und SRVR/SRVRP, Prospekt Nrn. 5.118.3/12.94/9.97, 5.117.3/2.96 und 5.116.0/8.97 Katalog 01 Rubrik 08, Sulzbach/Saar

Ivantysyn, J. u. M.: Hydrostatische Pumpen und Motoren, Würzburg: Vogel Verlag 1993

HYDAC: 2-Wege-Stromregelventile SRE, SR5E und SRVR/SRVRP, Prospekt Nrn. 5.118.3/12.94/9.97, 5.117.3/2.96 und 5.116.0/8.97 Katalog 01 Rubrik 08, Sulzbach/Saar

Ivantysyn, J. u. M.: Hydrostatische Pumpen und Motoren, Würzburg: Vogel Verlag 1993

Kalide, W.: Einführung in die technische Strömungslehre, München: Carl Hanser Verlag 1965

Kirst, T.: Hydraulik Fluidtechnik, Würzburg: Vogel Verlag 1991

Kirst, T.: Hydraulik Pneumatik Fluidik/Pneulogik, Darmstadt: Hoppenstedt Technik Tabellen Verlag 1991

Kracht: Zahnrad-Förderpumpen, Prospektkennziffer KF3-6.d.10.99, Werdohl

Mannesmann Rexroth: Grundlagen und Komponenten der Fluidtechnik Hydraulik, Der Hydraulik Trainer Band 1 (RD 00290/10.91), Lohr am Main: 1991

Mannesmann Rexroth: Proportional- und Servoventiltechnik, Der Hydraulik Trainer Band 2 (RD 00291/12.89), Lohr am Main: 1988

Mannesmann Rexroth: Projektierung und Konstruktion von Hydroanlagen, Der Hydrauliktrainer Band 3 (RD 00281/10.88), Lohr am Main: 1988

Mannesmann Rexroth: Rückschlagventil-Einbausatz Typ M-SR, Serie 1X, Druckschrift RD 20 380/04.92, Lohr am Main

Mannesmann Rexroth: Hydrozylinder Zugankerbauart, Nenndruck 70 bar,

Mannesmann Rexroth: Hydropumpen für die Antriebshydraulik, Katalog RD 00 190, Lohr am Main

Mannesmann Rexroth: Hydromotoren für die Antriebshydraulik, Katalog RD 00 195, Lohr am Main

Mannesmann Rexroth: Stetigventile, Regelungssysteme, Elektronik-Komponenten, Band 1: Stetigventile und Zubehör, Regelungssysteme, Katalog RD 00 155-01, Lohr am Main

Matthies, H. J.: Einführung in die Ölhydraulik, , Stuttgart: B. G. Teubner 1991

Murrenhoff, H.: Servohydraulik (Umdruck zur Vorlesung), Institut für fluidtechnische Antriebe und Steuerungen der RWTH Aachen, Aachen: Verlag Mainz 1998

Panzer, P., Beitler, G.: Arbeitsbuch der Ölhydraulik, Projektierung und Betrieb, Mainz: Krausskopf-Verlag 1969

Paetzold, W., Hemming, W.: Hydraulik und Pneumatik, Konstanz: Verlag Christiani 1997

Parker Hannifin: Hydraulik-Zylinder Serie 2H mit Stufendämpfung zur Steigerung von Leistung und Produktivität, Katalogkennziffer 1110 D, Kaarst

Parker Hannifin: Kompakt-Hydrozylinder Baureihe HMI nach ISO 6020/2 (1991), Baureihe HMD nach DIN 24554, Katalogkennziffer 1150/5-D, Kaarst

Prandtl, L., Oswatitsch, K., Wieghardt, K.: Führer durch die Strömungslehre, Braunschweig/Wiesbaden: Friedrich Vieweg & Sohn Verlagsgesellschaft 1984

Thoma, J.: Ölhydraulik – Entwurf und Gestaltung hydrostatischer Bauteile und Anlagen, München: Carl Hanser Verlag 1970

Truckenbrodt, E.: Strömungsmechanik, Berlin, Heidelberg, New-York: Springer-Verlag 1968

Sauer-Sundstrand: Axialkolben-Verstellpumpen (Technische Information), Baureihe 90, Prospektkennziffer TI-SPV90-D 11/98 369 298B

Sauer-Sundstrand: Axialkolbenmotoren (Technische Information), Baureihe 40, Prospektkennziffer TI-SMF/SMV40D 08/96 369 983

Shell: Änderung von Viskosität, Volumen und Dichte durch Temperatur und Druck, Mitteilungen des Shell Technischen Dienstes, MTO 2/Dr. St., Hamburg

Shell: Schmierstoffe, Herstellung - Eigenschaften – Anwendung, W. H. Kara, Hamburg 1986

Sigloch, H.: Technische Fluidmechanik, Düsseldorf : VDI-Verlag 1991

Sigloch, H.: Strömungsmaschinen, Grundlagen und Anwendungen, München: Carl Hanser Verlag 1993

Storz: Hydro-Normzylinder, Baureihe ZBD 1001 mit und ohne Endlagendämpfung, Druckschrift 13 210, Tuttlingen

Storz: Hydro-Standardzylinder mit beidseitiger Kolbenstange, Baureihe ZG 1601, Druckschrift 27310, Tuttlingen

Thoma, J.: Ölhydraulik, Entwurf und Gestaltung hydrostatischer Bauteile und Anlagen, München: Carl Hanser Verlag 1970

Vickers: Verstellbare Axialkolbenpumpen für Industrie-Anwendungen, Produktreihen PVQ200 und PVH300, Prospektkennziffern 5014.00/D/0297/A und 5016.00/D/0598/A

Vickers: Leiselaufende Flügelzellenpumpen – Baureihe V, Prospektkennziffer D-2343

Wärmetechnische Arbeitsmappe: Herausgegeben vom Verein Deutscher Ingenieure VDI-Gesellschaft Energietechnik, Düsseldorf: VDI-Verlag 1980

WEH: Rückschlagventile TVR1 und TVR2, Prospektblätter 7/97 und 01/01, Illertissen

www.ingramcontent.com/pod-product-compliance
Lightning Source LLC
Chambersburg PA
CBHW081116240526
45470CB00020B/3135